Lecture Notes in Computer Science 5122

Commenced Publication in 1973
Founding and Former Series Editors:
Gerhard Goos, Juris Hartmanis, and Jan van Leeuwen

Editorial Board

David Hutchison
  Lancaster University, UK

Takeo Kanade
  Carnegie Mellon University, Pittsburgh, PA, USA

Josef Kittler
  University of Surrey, Guildford, UK

Jon M. Kleinberg
  Cornell University, Ithaca, NY, USA

Alfred Kobsa
  University of California, Irvine, CA, USA

Friedemann Mattern
  ETH Zurich, Switzerland

John C. Mitchell
  Stanford University, CA, USA

Moni Naor
  Weizmann Institute of Science, Rehovot, Israel

Oscar Nierstrasz
  University of Bern, Switzerland

C. Pandu Rangan
  Indian Institute of Technology, Madras, India

Bernhard Steffen
  University of Dortmund, Germany

Madhu Sudan
  Massachusetts Institute of Technology, MA, USA

Demetri Terzopoulos
  University of California, Los Angeles, CA, USA

Doug Tygar
  University of California, Berkeley, CA, USA

Gerhard Weikum
  Max-Planck Institute of Computer Science, Saarbruecken, Germany

Llorenç Cerdà-Alabern (Ed.)

Wireless Systems and Mobility in Next Generation Internet

4th International Workshop
of the EuroNGI/EuroFGI Network of Excellence
Barcelona, Spain, January 16-18, 2008
Revised Selected Papers

 Springer

Volume Editor

Llorenç Cerdà-Alabern
Universitat Politècnica de Catalunya
Departament d'Arquitectura de Computadors
Jordi Girona 1-3, 08034 Barcelona, Spain
E-mail: llorenc@ac.upc.edu

Library of Congress Control Number: 2008938140

CR Subject Classification (1998): C.2, H.4, H.3

LNCS Sublibrary: SL 5 – Computer Communication Networks
and Telecommunications

ISSN	0302-9743
ISBN-10	3-540-89182-X Springer Berlin Heidelberg New York
ISBN-13	978-3-540-89182-6 Springer Berlin Heidelberg New York

Springer is a part of Springer Science+Business Media

springer.com

© Springer-Verlag Berlin Heidelberg 2008
Printed in Germany

Typesetting: Camera-ready by author, data conversion by Scientific Publishing Services, Chennai, India
Printed on acid-free paper   SPIN: 12440313   06/3180   5 4 3 2 1 0

Preface

The Network of Excellence on Next-Generation Internet[1] (EuroNGI), and its extension, Future-Generation Internet (EuroFGI), is a European project funded by the European Union within the IST Sixth Framework Program. The target of the EuroNGI/FGI network is to put together European research centers in the field of networking, focusing on next-generation Internet design and engineering. In all, 60 partners from technical universities and research centers in Europe participate in EuroNGI/FGI.

One of the "Integration Activities" within EuroNGI/FGI consists of the organization of an annual Workshop on Wireless and Mobility. The workshop intends to bring together leading researchers from the Network of Excellence in this field of research in order to identify the fundamental challenges and future perspectives of this important area.

This post-proceedings volume contains a selection of the research contributions presented at the fourth edition of the workshop. The workshop was held January 16–18, 2008 in Barcelona, Spain, organized by CompNet Research Group[2] from the Universitat Politècnica de Catalunya (UPC). The previous editions of the workshop were also organized by UPC in Sitges, Spain, in 2006; by Politecnico di Milano in Lago di Como, Italy, in 2005; and by Rheinisch-Westfälischen Technischen Hochschule Aachen in Dagstuhl, Germany, in 2004.

To participate in the workshop, the authors submitted an extended abstract of their ongoing work in the field of wireless and mobility. From a total of 39 submissions, 28 were accepted for presentation. Following the discussions and suggestions made in the workshop, the authors submitted 26 full papers that were carefully reviewed by experts on the field. Finally, 16 papers incorporating the numerous referee comments were accepted for publication in this book.

The papers contained in this book provide a general view of the ongoing research on wireless and mobility in the EuroNGI/FGI Network of Excellence, and they give a representative example of the problems currently investigated in this area.

May 2008   Llorenç Cerdà-Alabern

---

[1] http://www.eurongi.org
[2] http://research.ac.upc.edu/CompNet

Organization

Organization Committee

Jorge García-Vidal, Universitat Politècnica de Catalunya (UPC), Spain
Llorenç Cerdà-Alabern, Universitat Politècnica de Catalunya (UPC), Spain
José-M. Barceló-Ordinas, Universitat Politècnica de Catalunya (UPC), Spain
Manel Guerrero-Zapata, Universitat Politècnica de Catalunya (UPC), Spain

Referees

Sponsoring Institutions

EU-funded Euro-NGI/FGI Network of Excellence
Universitat Politècnica de Catalunya

Table of Contents

Cellular Networks

Performance Evaluation of DTSN in Wireless Sensor Networks

Francisco Rocha[1], António Grilo[1,2], Paulo Rogério Pereira[1,2],
Mário Serafim Nunes[1,2], and Augusto Casaca[1,2]

[1] INESC, Rua Alves Redol, nº 9, Lisboa, Portugal
[2] IST, Av. Rovisco Pais, Lisboa, Portugal
rocha.francisco@gmail.com, antonio.grilo@inov.pt, prbp@inesc.pt,
mario.nunes@inov.pt, augusto.casaca@inesc.pt

Abstract. The guaranteed delivery of critical data is an essential requirement in most Wireless Sensor Network (WSN) applications. The paucity of energy, communication, processing and storage resources in each WSN node causes the TCP transport model (widely used in broadband networks) to be inefficient in WSNs, a reason why new WSN-specific reliable transport protocols have been proposed in the past few years. This paper presents one of these protocols, the Distributed Transport for Sensor Networks (DTSN). DTSN is able to efficiently support unicast communications in WSNs due to its capabilities to tightly control the amount of signaling and retransmission overhead. The basic loss recovery algorithm is based on Selective Repeat ARQ, employing both positive and negative acknowledgements. Caching at intermediate nodes is used to avoid the inefficiency typical of the strictly end-to-end transport reliability commonly assumed in broadband networks. DTSN is currently implemented in TinyOS. Preliminary simulation results using this code show that DTSN is quite efficient providing block oriented reliability, while the caching mechanism employed in DTSN decreases packet delay for more than one hop.

Keywords: Energy-efficiency, Reliable Data Transport, Quality of Service, Wireless Sensor Networks.

1 Introduction

The guaranteed delivery of critical data is an essential requirement in most Wireless Sensor Network (WSN) applications. Illustrative examples are: battlefield surveillance, intrusion detection and E-health monitoring applications, where critical alerts must be timely and reliably delivered to the monitoring stations that act on those data; and industrial control applications, where commands must be timely and reliably delivered to the actuators (e.g., robotic arm). Moreover, while in the beginning WSNs were thought to convey very simple data such as alerts, actuator commands and physical measurements, the recent availability of low-cost miniaturized hardware such as CMOS cameras and microphones have led to the possibility of transmission of larger data blocks filled with sound samples and/or images. Given the paucity of resources and high bit error rate (BER) values typically

L. Cerdà-Alabern (Ed.): Wireless and Mobility 2008, LNCS 5122, pp. 1–9, 2008.

featured by these networks, bulky data delivery can only be effected if the bandwidth utilization is maximized. This entails minimizing the end-to-end packet loss ratio (PLR), which is a traditional responsibility of the transport function.

Proven transport protocols like TCP [1], designed to support user applications in infrastructure networks, usually present significant inefficiencies when employed without considerable modification in WSN systems. Several proposals have been made for alternative reliable transport protocols, but only a few of them are suited to support block-oriented data delivery. The shortcomings found in these existent proposals led to the development of the Distributed Transport for Sensor Networks (DTSN) protocol [2], which is based on the following principles:

- Reliable transmission of block-oriented data;
- Energy-efficiency: to avoid useless wasting of energy resources through minimization of the control and retransmission overhead;
- Distributed Functionality: to exploit the WSN storage resources in a cooperative way, so that the scalability of WSN operation in a multihop environment is increased.

This paper presents a performance evaluation of DTSN based on its TinyOS [3] implementation.

The rest of the paper is structured as follows. Section 2 presents the most relevant related work; section 3 describes the DTSN protocol; section 4 presents the simulation results; section 5 concludes the paper and lists some topics of ongoing and future work.

2 Related Work

There are several proposals of reliable transport protocols for WSNs. This section only covers the proposals for block-oriented reliable transport that were considered more relevant for the development of DTSN.

Reliable Multi-Segment Transport (RMST) [4] was designed to work on top of the Directed Diffusion [5]. It offers two simple services: data segmentation/reassembly and guaranteed delivery. RMST can operate end-to-end or in a store-and-forward mode where intermediate nodes care to receive all the fragments of a block before forwarding it. RMST is affected by various weaknesses that make it somewhat unreliable and inefficient. First of all, there is no full guarantee of delivery. Since it only uses negative acknowledgements (NACKs) for loss recovery, when none of the fragments of a data block are received by an intermediate or destination node, the loss can not be detected because the sink is not aware of the transmission and there is no end-to-end positive ACK to finalize the transaction.

Distributed TCP Caching (DTC) [6] is a TCP enhancement to make it more efficient in WSNs. It improves the transmission efficiency by compressing the headers and the use of caching at selected nodes in the path from source to destination, thus minimizing end-to-end retransmissions. DTC is fully compatible with TCP, leaving the endpoints of communication unchanged – it only requires changes in the logic of intermediate nodes. The use of caching significantly improves the efficiency of packet loss recovery, minimizing the energy spent with retransmissions. However, like in TCP, DTC features an inefficiency associated with the transmission of unnecessary positive ACKs, which is totally controlled by the receiver.

Pump Slow Fetch Quickly (PSFQ) [7] is a protocol primarily designed to offer downstream multicast guaranteed delivery for dynamic code update, though it can also be used for upstream unicast communication. It supports the reliable dissemination of consistent data blocks along the network within a user defined scope. Its behavior fluctuates between multihop forwarding during the data dissemination phase (Pump phase), and store and forward when losses are detected. Upon loss detection, forwarding is stopped and recovery phase (Fetch phase) is started. Data is disseminated through broadcasting during a pre-defined time interval at the end of which flow control is applied. That approach allows to stop the propagation of gaps in the fragment sequence as soon as possible and to quickly recover missing fragments from neighbors. Every node, in fact, stores the received data to be able to provide missing fragments to requesting neighbors. The use of flow control at every hop introduces a high delay in the communication. Besides, the fact that PSFQ only employs NACK feedback (like RMST) causes it to be unable to detect the loss of all fragments of a block at once. Regarding intermediate caching, data is reconstructed at each hop. While this makes sense for dynamic code update (i.e., each node must get all the executable code fragments), that can be very limiting for other applications (audio or image transmission) since it poses significant requirements on node storage capabilities.

3 Distributed Transport for Sensor Networks

The DTSN specification [2] can be divided into a full reliability service, and a differentiated reliability service. The latter is based on the former and is able to support several reliability grades based on the integration of partial buffering at the source, intermediate node caching and erasure coding. This paper focuses on the full reliability service.

The full reliability service was thought for critical data transfer requiring end-to-end full reliability. It is achieved by a Selective Repeat Automatic Repeat reQuest (ARQ) mechanism that uses both negative acknowledgement (NACK) and positive acknowledgement (ACK) packets. Soft requirements of routing path stability and bi-directionality are placed on the routing layer in order to leverage the intermediate node optional cache mechanism of DTSN for performance improvement, although DTSN is able to operate end-to-end as well.

In DTSN, a session is a source/destination relationship univocally identified by the tuple <source address, destination address, application identifier, session number>, designated the session identifier. The session is soft-state by nature both at the source and the destination, being created when the first packet is processed and terminated upon the expiration of an activity timer (provided that no activity is detected and there are no pending delivery confirmations). A randomly chosen session number is appended in order to unambiguously distinguish between successive sessions sharing the endpoint addresses and application identifier. Within a session, packets are sequentially numbered. The Acknowledgement Window (AW) is defined as the number of packets that the source transmits before generating a confirmation request (Explicit Acknowledgement Request – EAR). The output buffer at the sender works as a sliding window, which can span more than one AW. The size of the output buffer and of the AW depend on the specific scenario, namely on the memory constraints of individual nodes.

The DTSN session management algorithm at the source works as follows. The source transmits each packet coming from the higher layers, storing it in the output buffer, so that it can be retransmitted latter if required. Upon the transmission of each set of packets equal to the AW size, or when the output buffer is full, or when the higher layer protocols have not sent any data during a predefined timeout period, the source requests a delivery confirmation message from the destination by means of an EAR. This may take the form of a bit flag piggybacked in the last data packet (e.g., confirmation request due to the AW size being reached or the output buffer becoming full) or an independent packet (e.g., confirmation request due to the expiration of the EAR timer). Each time a confirmation message (either ACK or NACK) is received, the source frees the output buffer entries whose delivery is confirmed. The reception of an ACK means that there are no gaps in the sequence of packets sent before the respective EAR. On the other hand, a NACK includes a bitmap, where each bit represents a different sequence number (starting from a base sequence number indicated in the packet header) and indicates whether the corresponding packet was correctly received or not. Its reception causes the source to retransmit the data packets that were not delivered successfully. An EAR is sent after retransmission, which may be piggybacked in the last of the retransmitted packets. After sending an EAR, the source launches an EAR timer. If the EAR timer expires before an ACK/NACK is received, the source retransmits the EAR packet.

The DTSN algorithm at the destination works as follows. Upon reception of a data packet with a new session identifier, a new session record is created. However, if a session already exists with the same addresses and application identifier, but the session number is different from the recorded one, the session record is reset and the new session number replaces the old one. The destination then collects the data packets that belong to that flow, delivering in-sequence packets to the higher layer protocol. Upon reception of an EAR, the destination sends an ACK or NACK depending on the existence of gaps in the received data packet stream. Upon the expiration of an activity timer, the session record is deleted and the higher layer protocol is notified in case there are unconfirmed packets.

As already explained, the strict end-to-end transport reliability model used in TCP is not suited for WSNs because it leads to extra consumption of the scarce bandwidth and energy resources. This is caused by the fact that missing packets (and some control packets like NACKs) are retransmitted end-to-end, expending bandwidth and energy in all links/nodes in the path between source and destination. Caching at intermediate nodes is the mechanism employed by DTSN to counter this inefficiency. In DTSN, each node keeps a cache of intercepted packets, managed according to a suitable replacement policy (currently it is FIFO). The packets are stored in cache with probability p, and may belong to different sessions whose end-to-end routing path includes the node in question. Each time an intermediate node receives a NACK packet, it analyses its body and searches for corresponding data packets that are missing at the destination. In case a missing packet is detected, the intermediate node

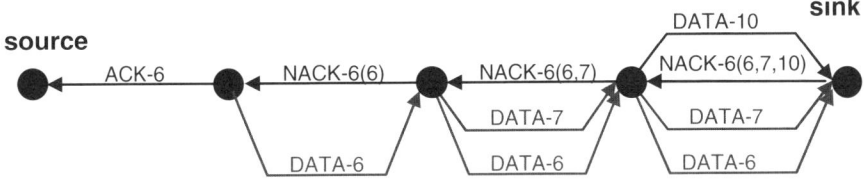

Fig. 1. Example of cache operation and NACK filtering at intermediate nodes

retransmits it. It also changes the NACK contents before resending it, modifying the bitmap so that its retransmitted packets are not included (eventually, the NACK may become an ACK if all missing packets were found in cache). In this way, the source will only have to retransmit those data packets that were not cached at intermediate nodes, decreasing the average hop length of the paths followed by retransmitted packets. Intermediate nodes also discard confirmed packets from the cache and process the session number header field in data packets, replacing any cache entries whose session number is outdated. Fig. 1 illustrates the operation of the cache mechanism for a source-sink path with three intermediate nodes. In this example, the sink node sends a NACK packet stating it is expecting packet 6, so it has received all packets up to packet 5, and packets 6, 7 and 10 are in the missing packet list. The previous node has packet 10, so it sends it to the sink and retransmits the NACK back, removing packet 10 from the missing packets list. The next node has packet 7. The following node has packet 6. As there are no more missing packets, the NACK is turned into an ACK that reaches the source.

The caching probability p at the intermediate nodes should be defined taking into account various factors like cache size, traffic load and EAR frequency. The value of p must be chosen to maximize the probability of cache hit for the NACK requested data along the path. If several nodes have a missing packet cached that is simultaneously retransmitted when they receive a NACK, the resulting collision will degrade performance. To reduce this effect, only nodes in the end-to-end path cache packets in the current DTSN implementation. Ideally, the packets should be cached as close as possible to the destination to reduce the number of radio transmissions necessary to recover from a missing packet. So, the caching probability should decrease with the hop distance to the destination. Naturally, this requires that the routing layer gives this information to the transport layer. To save cache memory, different nodes should cache different packets; consequently the caching probability should be less than 1.

4   Simulation Results

Simulations of the TinyOS 2.x implementation of DTSN were conducted using the TOSSIM simulation environment.

The radiofrequency (RF) parameters of sensor nodes correspond to those of the MICAz motes developed by Crossbow [8], which support a bitrate of 250 kbps based on the IEEE 802.15.4 standard for the MAC and physical layers [9]. The simulations were made considering a linear topology of evenly spaced sensor nodes, where the transmission power is 0 dBm (maximum power for MICAz nodes), the path loss between each pair of nodes is 70 dB (this corresponds to approximately 10 meters of distance assuming a log-distance path loss model with a distance exponent of 3) and the background noise level is -110 dBm.

Destination-Sequenced Distance-Vector routing (DSDV) [10] is used as the routing protocol underlying DTSN. A continuous traffic pattern was generated at the first node in the line, with the last node as the destination. The data payload size is 2 bytes, adding to the 8 bytes of DTSN overhead, 5 bytes of DSDV overhead and 13 bytes of MAC overhead. The 8 bytes of DTSN overhead include 1 byte for flags (e.g., ACK, NACK), 1 byte for the packet sequence number and 6 bytes for the session identifier. Some overhead may be reduced by combining information of source and destination addresses from the routing layer and transport layers. As a matter of fact, the source and destination addresses could be retrieved from the DSDV routing, saving 4 bytes. Intermediate nodes retransmit cached packets without modification, being transparent to the transport session. An additional byte may be saved by sending the application identifier in the TinyOS Active Message (AM) type field. As these optimizations require cross layer interactions, they were not used in the experiments.

The DTSN transmission window size was configured as 50 packets, with two AWs of 25 packets each. The EAR timeout was set as 250 ms. The maximum number of EAR timeouts is 10. The session activity timeout is 7.5 s at the sender and 10 s at the receiver. The caching probability p (when used) is set to 1 and the cache size is enough for 50 packets. This caching configuration should maximize the cache hit ratio. As a disadvantage, it will also maximize memory consumption. Exhaustive tests with different parameters to save memory were left for future work.

The results allow a comparison between raw DSDV transmission (without reliable transport on top) and DTSN with and without the intermediate node caching mechanism over DSDV routing. Each point in the graphics corresponds to an average over ten independent runs, each consisting of the transmission of 1000 packets back-to-back (subject to local interlayer flow control). The achieved packet loss ratio, user data throughput, average delay and per-packet overhead (measured as the total number of radio transmissions per successfully delivered packet received at the destination) are depicted respectively in Fig. 2, Fig. 3, Fig. 4 and Fig. 5 as a function of the hop distance between source and destination.

The packet loss ratio for DSDV increases quite steadily with the hop distance as noise and unrecovered collisions cause corrupted packets to be lost. This justifies the use of a reliable transport protocol like DTSN that achieves zero packet loss, as expected. On the other hand, DSDV achieves higher throughput and much lower delay as compared with DTSN, which mimics the difference between TCP and UDP in IP networks. The DSDV throughput is naturally higher, since packets have smaller headers. The DTSN delay is much larger than the DSDV delay, as lost packets are retransmitted by DTSN after a timeout. Using an adaptive timeout delay mechanism as used in the TCP protocol may reduce the delay experienced by packets. This investigation was left for further work.

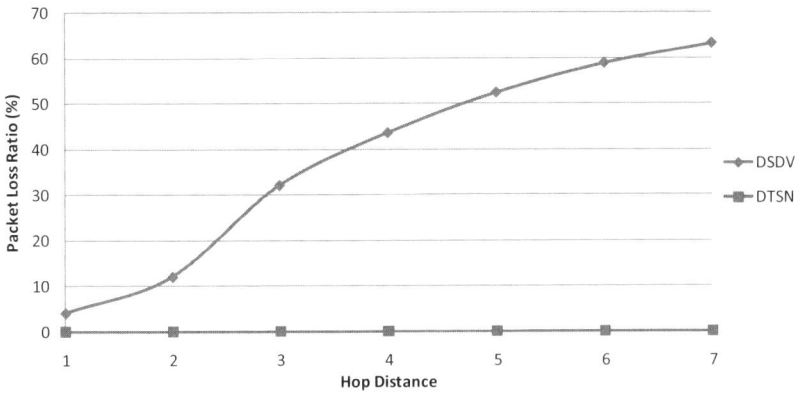

Fig. 2. Packet Loss Ratio

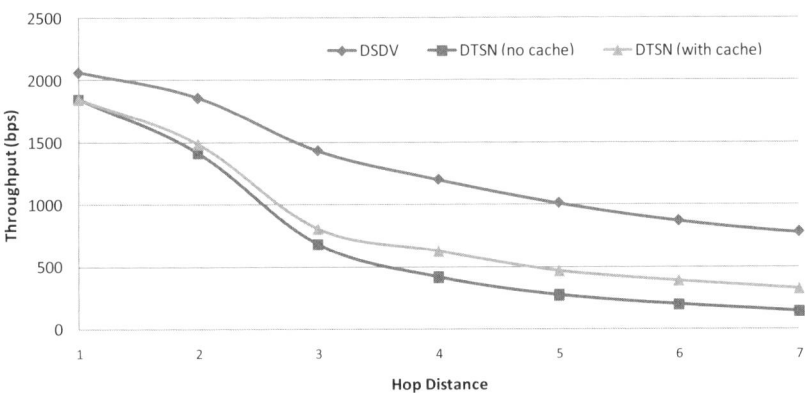

Fig. 3. User data throughput considering 2 data bytes per packet

Regarding the number of radio transmissions per packet shown in Fig. 5, it can be observed that this value increases with the hop distance for all cases. Even though DSDV does not do any retransmission, collisions may trigger MAC layer retransmissions. For this reason, the number of radio transmissions is larger than the hop distance. Although DSDV clearly presents a lower number of radio transmissions as compared with DTSN without cache, the difference is not very significant when compared with DTSN with the cache mechanism turned on. For any hop distance greater than 1, the cache mechanism is used and effectively reduces the number of radio transmissions, saving energy. For hop distances greater than 5, DTSN with cache has even less radio transmissions than DSDV. This result can be justified as fewer packets are effectively delivered in DSDV when the number of hops increases, resulting in increased wasted radio transmissions.

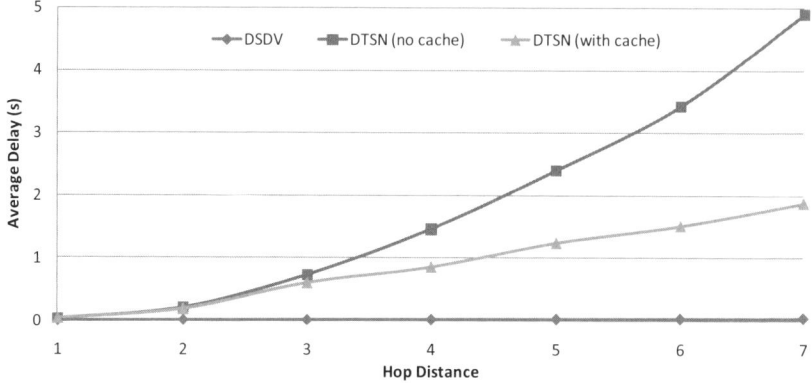

Fig. 4. Average Delay

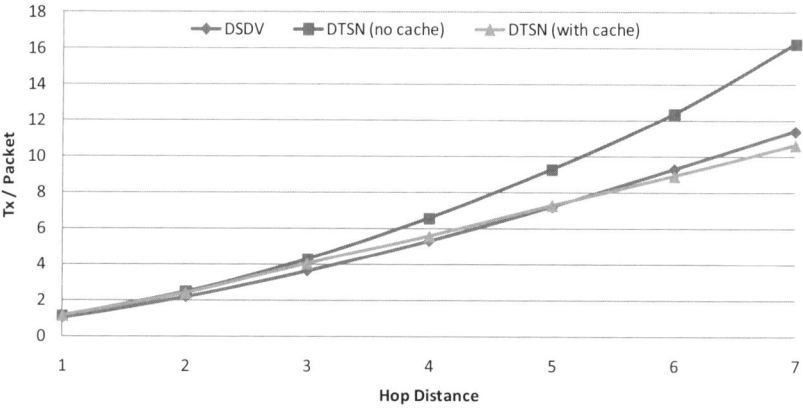

Fig. 5. Average number of RF transmissions per successfully delivered data packet

5 Conclusion

This paper has presented a performance evaluation of the Distributed Transport for Sensor Networks (DTSN). This reliable transport protocol for wireless sensor networks minimizes control overhead by placing the control of the session at the sender, which explicitly requests the data delivery confirmations from the receiver. The use of NACK packets with indication of missing data packets enhances performance. A caching mechanism at the intermediate nodes attempts to minimize end-to-end retransmissions.

The performance evaluation of DTSN was based on a TinyOS 2.x implementation over the DSDV routing protocol. The simulation results attest the effectiveness of DTSN to minimize the number of RF transmissions per packet. The use of the caching mechanism proved to be effective, leveling the energy performance of DTSN using cache with the performance of the underlying DSDV, with the advantage of having a reliable protocol.

Future work will focus on the evaluation of DTSN for different packet sizes, caching policies, network topologies and traffic load values. Additional evaluations will compare DTSN with other WSN transport protocols and test it with other routing and MAC protocols. Additionally, future work will also include the implementation and evaluation of the differentiated reliability service for Wireless Multimedia Sensor Networks and of an adaptive timeout configuration mechanism.

Acknowledgments. The authors acknowledge the research funding from IST FP6 UbiSec&Sens project and IST FP6 EuroFGI's E2E-WSN specific research project.

References

1. Postel, J.: Transmission Control Protocol, RFC 793, IETF (September 1981)
2. Marchi, B., Grilo, A., Nunes, M.: DTSN – Distributed Transport for Sensor Networks. In: Proceedings of the IEEE Symposium on Computers and Communications (ISCC 2007), Aveiro, Portugal (2007)
3. Hill, J., Szewczyk, R., Woo, A., Hollar, S., Culler, D., Pister, K.: System Architecture Directions for Networked Sensors. In: Proceedings of the Ninth International Conference on Architectural Support for Programming Languages and Operating Systems (ASPLOS) (November 2000)
4. Stann, F., Heidemann, J.: RMST: Reliable Data Transport in Sensor Networks. In: Proceeding of the 1st IEEE International Workshop on Sensor Net Protocols and Applications, May 2003, pp. 1–11. IEEE, Anchorage, Alaska (2003)
5. Intanagonwiwat, C., Govindan, R., Estrin, D.: Directed Diffusion: A Scalable and Robust Communication Paradigm for Sensor Networks. In: Proceedings of the Sixth Annual International Conference on Mobile Computing and Networks (MobiCON) (August 2000)
6. Dunkels, A., Alonso, J., Voigt, T., Ritter, H.: Distributed TCP Caching for Wireless Sensor Networks. In: Proceedings of the 3rd Annual Mediterranean Ad-Hoc Networks Workshop, Bodrum, Turkey (June 2004)
7. Wan, C.-Y., Campbell, A., Krishnamurthy, L.: PSFQ: A Reliable Transport Protocol for Wireless Sensor Networks. In: Proceedings of WSNA 2002, Atlanta, Georgia, USA (2002)
8. MICAz datasheet, Crossbow (24/4/2007), http://www.xbow.com/Products/productdetails.aspx?sid=164
9. IEEE Std. 802.15.4, Wireless Medium Access Control (MAC) and Physical Layer (PHY) Specifications for Low-Rate Wireless Personal Area Networks (LR-WPANs) (2003)
10. Perkins, C., Bhagwat, P.: Highly Dynamic Destination-sequenced Distance-vector routing (DSDV) for Mobile Computers. In: SIGCOMM, pp. 234–244 (1994)

Diffusion Approximation Model for the Distribution of Packet Travel Time at Sensor Networks

Tadeusz Czachórski[1], Krzysztof Grochla[1], and Ferhan Pekergin[2]

[1] Institute of Theoretical and Applied Informatics
Polish Academy of Sciences
Baltycka 5, 44–100 Gliwice, Poland
{tadek,kil}@iitis.gliwice.pl
[2] LIPN, Université Paris-Nord, 93430 Villetaneuse, France
pekergin@lipn.univ-paris13.fr

Abstract. We propose a model to estimate the probability density function of the distribution of a packet travel time in a multihop wireless sensor network. The model is based on diffusion approximation and it takes into consideration the heterogeneity of the propagation medium and of the distribution of relay nodes. The modeled system parameters, e.g. the packet loss probability or the network topology, may depend on time.

Keywords: diffusion approximation, transient analysis, wireless networks, sensor networks.

1 Introduction

Prediction of a packet travel time in wireless sensor networks is still an open issue. A sensor network, see e.g. [1] consist of a large number of simple nodes scattered randomly over a certain area, having ability to route packets to their neighbors and finally to the sink which collects the data sent to it via multihop transmission.

The topology of such networks is in most of cases uncertain and it changes in time (due to nodes movement or failures), hence special routing algorithms were proposed, e.g. [7], [8] to face this situation. As it is also hard to introduce global addressing, the routing decision must be made without the complete information about the network. We consider the same network model as in [6] – a packet wireless network in which nodes are distributed over an area, but where we do not know about the presence, exact location, or reliability of nodes. The packets are forwarded to a node which is most probably nearer to the destination, but it is also possible that a transmission may actually move the packet further away from the sink or send it to a node which is in the same distance to the destination (see for example [10]). It may also happen that a packet cannot be forwarded any further, that the intermediate node has a failure, or that the packet is lost through noise or some other transient effect. In that case, the packet may be retransmitted after a time-out period has elapsed, either by the source or from some intermediate storage location on the path which it traversed before it was lost.

L. Cerdà-Alabern (Ed.): Wireless and Mobility 2008, LNCS 5122, pp. 10–25, 2008.

2   Model Formulation

Recently Gelenbe [6] proposed a model based on diffusion approximation to estimate the mean transmission time from a source to destination in a random multihop medium. In this model a value of the diffusion process represents the distance defined as the number of hops between the transmitted packet and its destination (sink). Due to complex topology and transmission constraints, it is not sure that each one-hop transmission makes this distance shorter and the changes of the distance may be considered as random process. This justifies the use of diffusion process to characterize it. Diffusion approximation is a classical method used in queueing theory to represent a queue length or queueing time e.g. [3], in case of general independent distributions of interarrival and service times. Diffusion process is a continuous stochastic process but it is used to approximate some discrete processes, see [2], like – as mentioned above – the number of customers in a queue; here it represents the number of hops still remaining to the packet's destination.

If $N(t)$ denotes the number of hops remaining to destination at time t, we construct a diffusion process $X(t)$ such that its density function $f(x, t; x_0)$ approximates probability distribution $p(n, t; n_0)$ of the process $N(t)$, $N(0) = n_0$: $f(n, t; n_0) \approx p(n, t; n_0)$. The density function $f(x, t; x_0)$

$$f(x, t; x_0)dx = P[x \leq X(t) < x + dx \mid X(0) = x_0]$$

is defined by the diffusion equation

$$\frac{\partial f(x, t; x_0)}{\partial t} = \frac{\alpha}{2} \frac{\partial^2 f(x, t; x_0)}{\partial x^2} - \beta \frac{\partial f(x, t; x_0)}{\partial x}, \tag{1}$$

where the parameters β and α define respectively the mean and variance of infinitesimal changes of the diffusion process. To maintain them similar to the considered process $N(t)$, they should be chosen as

$$\beta = \lim_{\Delta t \to 0} \frac{E[N(t + \Delta t) - N(t)]}{\Delta t},$$

$$\alpha = \lim_{\Delta t \to 0} \frac{E[(N(t + \Delta t) - N(t))^2] - (E[N(t + \Delta t) - N(t)])^2}{\Delta t}.$$

In general, the parameters may depend on time and on the current value of the process, $\beta = \beta(x, t)$ and $\alpha = \alpha(x, t)$, as the propagation medium and distribution of relay nodes may be heterogeneous in space and the system characteristics may change over time. We include this case in the proposed approach.

Gelenbe in [6] constructs an ergodic process going repetitively from starting point to zero and considers its steady-state properties. Here, to obtain the distribution (and not only the mean transmission time as given in [6]), we use transient solution of diffusion equation and we consider only a single process. Let us repeat that the process starts at $x_0 = N$ and ends when it successfully comes to the absorbing barrier at $x = 0$; the position x of the process corresponds to the current distance between the packet and its destination, counted in hops.

3   Model without Deadline and without Losses

In this simplest case we consider diffusion equation (1) with constant coefficients, supplemented with absorbing barrier at $x = 0$. This barrier is expressed by the boundary condition $\lim_{x \to 0} f(x, t; x_0) = 0$. The process starts at x_0: $X(0) = x_0$ and ends when it comes to the barrier. The diffusion process is defined at the interval $(0, \infty)$. Let us denote the solution of the diffusion equation in this case by $\phi(x, t; x_0)$; it is obtained using mirror method, see e.g. [2]

$$\phi(x, t; x_0) = \frac{1}{\sqrt{2\Pi \alpha t}} \left[e^{-\frac{(x_0 - x - |\beta| t)^2}{2\alpha t}} - e^{\frac{2\beta x_0}{\alpha}} e^{-\frac{(2x_0 - |\beta| t)^2}{2\alpha t}} \right]. \tag{2}$$

Fig. 1 presents a plot of the function $\phi(x, t; x_0)$. The function allows us to determine the speed of the changes of the probability mass being still on the axis x outside of the barrier, hence the density of the probability that the process ends at time t. This is the density $\gamma_{x_0, 0}(t)$ of the first passage time from $x = x_0$ to $x = 0$, i.e. an estimation of the density of a packet transmission time through x_0 hops from an initial node to the sink:

$$\gamma_{x_0, 0}(t) = \frac{\partial}{\partial t} \int_{0+}^{\infty} f(x, t : x_0) dx = \int_{0+}^{\infty} \frac{\partial f(x, t : x_0)}{\partial t} dx =$$

$$= \int_{0+}^{\infty} \left[\frac{\alpha}{2} \frac{\partial^2 f(x, t; x_0)}{\partial x^2} - \beta \frac{\partial f(x, t; x_0)}{\partial x} \right] dx =$$

$$= \lim_{x \to 0} \left[\frac{\alpha}{2} \frac{\partial}{\partial x} \phi(x, t; x_0) - \beta \phi(x, t; x_0) \right] = \frac{x_0}{\sqrt{2\Pi \alpha t^3}} e^{-\frac{(x_0 - |\beta| t)^2}{2\alpha t}}. \tag{3}$$

In this derivation we replace the left side of the diffusion equation (1) by its right side and we assume that the function $f(x, t : x_0)$ and its first and second partial

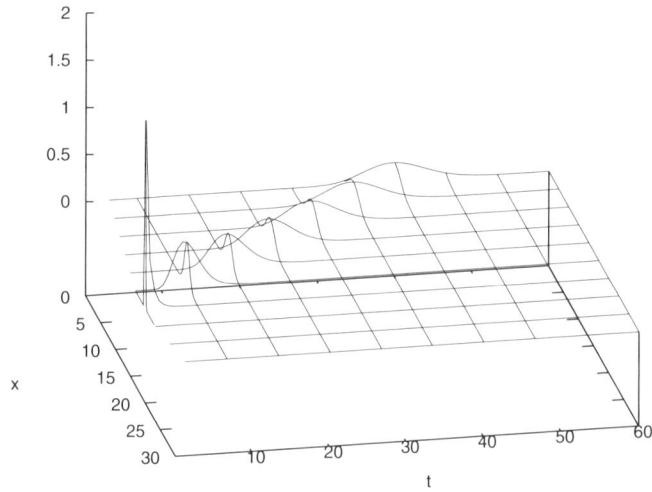

Fig. 1. Density function $\phi(x, t; x_0)$ of the diffusion process with absorbing barrier, $x_0 = 20$, $\alpha = 0.1$, $\beta = -0.5$

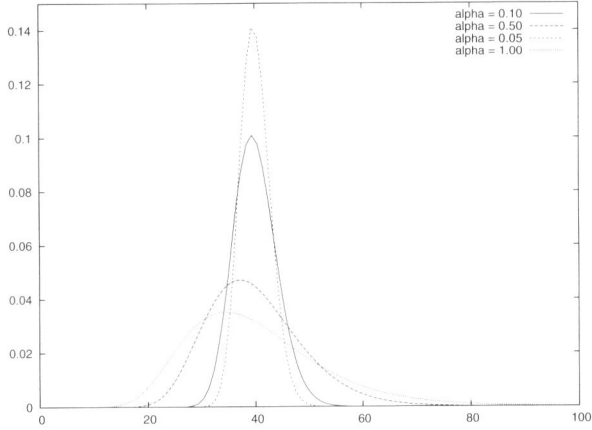

Fig. 2. Distribution of first passage time from x_0 to 0, $\gamma_{x_0,0}(t)$, $x_0 = 20$, $\beta = -0.5$, $\alpha = 0.05, 0.1, 0.5, 1.0$

derivatives over x tend to zero as x tends to infinity. Some exemplary curves of $\gamma_{x_0,0}(t)$ are presented in Fig. 2.

4   Introduction of the Deadline

Denote by τ the time after which a packet is considered lost and is retransmitted by the source. Knowing the density $\gamma_{x_0,0}(t)$ of the travel time from x_0 to 0, we can determine the probability $p_\tau = \int_\tau^\infty \gamma_{x_0,0}(t)dt$ that a packet at the moment τ is still on its way, see Fig. 3. In the model, at $t = \tau$ we shift this probability mass p_τ to x_0 and we restart the diffusion process. We may of course introduce an additional delay before the restart.

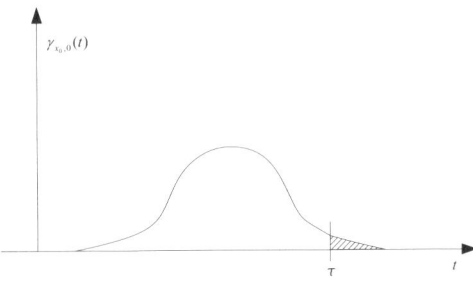

Fig. 3. Probability $p_\tau = \int_\tau^\infty \gamma_{x_0,0}(t)dt$ that a packet at the moment τ is still on its way

5   Modelling Heterogeneous Medium and Losses

To reflect the fact that the transmission conditions may be different for each hop, the diffusion interval is divided into unitary subintervals corresponding to single hops. The

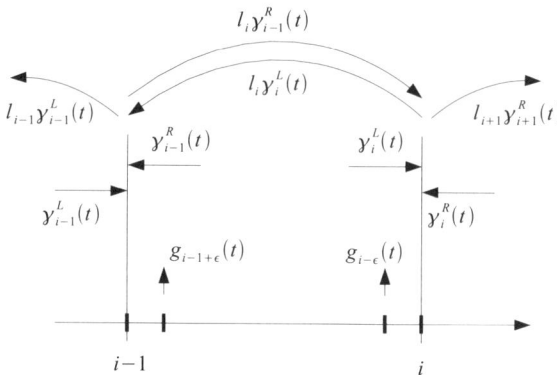

Fig. 4. Diagram of probability mass circulation due to non-homogenous diffusion parameters and due to losses with probability l_i at i-th hop, $i = 2, \ldots, D - 1$

subintervals are separated by fictive barriers allowing us to balance the probability density flows between them. We limit the whole interval to a value corresponding to the size of the network $x \in [0, D]$, the starting point x_0 is somewhere inside this interval. As in general $\beta < 0$ (i.e. a packet has a tendency of going towards the sink), the probability of reaching the right barrier by the diffusion process is small. If however the process reaches the right barrier, it is immediately sent to the point $x = D - \varepsilon$ and the process is continued.

An interval i, $x \in [i - 1, i]$ represents the packet transmission when it is i hops distant from the sink. We assume that parameters β_i, α_i are specific to this interval and we assume also the loss probability l_i within this interval.

When the process approaches one of the barriers limiting the subinterval i, for example the barrier at $x = i$, the barrier acts as an absorbing one, but then immediately the process reappears on the other side of the barrier with probability $(1 - l_i)$ (probability of successful transmission) or with probability l_i it jumps back to the node that it visited previously, i.e. to the barrier at $x = i + 1$ or at $x = i - 1$.

Let $\gamma_i^L(t)$ represent the flow coming to the barrier placed at $x = i$ from its left side and $\gamma_i^R(t)$ be the flow coming to this barrier from its right side. The flows start diffusion processes at both sides of the barrier, respectively $\gamma_i^R(t)$ reappears at $x = i - \varepsilon$ and $\gamma_i^L(t)$ at $x = i + \varepsilon$ but the intensities of the trespassing flows are reduced by flows corresponding to the loss of packet during the previous hop transmission. If we assume that the loss may be repaired by sending the lost packet from the neighboring node, the flow $\gamma_i^L(t)l_i$ is sent to $x = i - 1 + \varepsilon$ and the flow $\gamma_i^R(t)l_{i+1}$ is sent to $x = i + 1 - \varepsilon$. The circulation of probability mass for i-th interval, representing i-th hop, is presented in Fig. 4. Thus inside the interval i the process starts with intensities

$$g_{i-1+\varepsilon}(t) = (1 - l_{i-1})\gamma_{i-1}^L(t) + l_i\gamma_i^L(t)$$
$$g_{i-\varepsilon}(t) = (1 - l_{i+1})\gamma_i^R(t) + l_i\gamma_{i-1}^R(t)$$

where $g_{i-1+\varepsilon}(t)$ and $g_{i-\varepsilon}(t)$ are the probability densities that the diffusion process starts at time t at the point $x = i - 1 + \varepsilon$ and $x = i - \varepsilon$.

The flows $\gamma_{i-1}^L(t)$ and $\gamma_i^R(t)$ for the i-th interval are obtained as

$$\gamma_{i-1}^R(t) = \lim_{x \to (i-1)} \left[\frac{\alpha_i}{2} \frac{\partial f_i(x,t;\psi_i)}{\partial x} - \beta_i f_i(x,t;\psi_i) \right] = \lim_{x \to (i-1)} \left[\frac{\alpha_i}{2} \frac{\partial f_i(x,t;\psi_i)}{\partial x} \right],$$

$$\gamma_i^L(t) = - \lim_{x \to (i)} \left[\frac{\alpha_i}{2} \frac{\partial f_i(x,t;\psi_i)}{\partial x} - \beta_i f_i(x,t;\psi_i) \right] = - \lim_{x \to (i)} \left[\frac{\alpha_i}{2} \frac{\partial f_i(x,t;\psi_i)}{\partial x} \right].$$

where $f_i(x,t;\psi_i)$ is defined below, see Eq. (4). As the fictive barriers at the extremities of an interval i act as the absorbing one, the value of the function drops to zero at $x = i - 1$ and $x = i$, at the end of computations we put at $x = i$ the mean of $f_i(i - \varepsilon, t; \psi_i)$ and $f_{i+1}(i + \varepsilon, t; \psi_{i+1})$.

If the lost packets are retransmitted with a certain delay, e.g. after a random time distributed with density function $l(t)$ [if this time is constant and equal r then $l(t) = \delta(t - r)$], we rewrite the above equations as

$$g_{i-1+\varepsilon}(t) = (1 - l_{i-1})\gamma_{i-1}^L(t) + l_i \gamma_i^L(t) * l(t)$$

$$g_{i-\varepsilon}(t) = (1 - l_{i+1})\gamma_i^R(t) + l_i \gamma_{i-1}^R(t) * l(t)$$

where $*$ denotes the operation of convolution.

Within each subinterval we have diffusion process with two absorbing barriers, e.g. for i-th interval at $x = i - 1$ and $x = i$ and with two points when the process is started, at $i - 1 + \varepsilon$ with intensity $g_{i-1+\varepsilon}(t)$ and at $i - \varepsilon$ with intensity $g_{i-\varepsilon}(t)$.

The density of the diffusion process started at x_0 within an interval $(0, D)$ having the absorbing barriers at $x = 0$ and $x = D$ has the form, see [2]

$$\phi(x,t;x_0) = \begin{cases} \delta(x - x_0), t = 0 \\ \dfrac{1}{\sqrt{2\Pi\alpha t}} \displaystyle\sum_{n=-\infty}^{\infty} \left\{ \exp\left[\dfrac{\beta x_n'}{\alpha} - \dfrac{(x - x_0 - x_n' - \beta t)^2}{2\alpha t} \right] \right. \\ \left. - \exp\left[\dfrac{\beta x_n''}{\alpha} - \dfrac{(x - x_0 - x_n'' - \beta t)^2}{2\alpha t} \right] \right\}, t > 0 , \end{cases}$$

where $x_n' = 2nD$, $x_n'' = -2x_0 - x_n'$.

Let $f_i(x,t;\psi_i)$ denote the density of the overall diffusion process with parameters β_i, α_i placed at the interval i where $x \in (i - 1, i)$, being continuously started at points $x = i - 1 + \varepsilon$ and $x = i - \varepsilon$ with densities $g_{i-1+\varepsilon}(t)$ and $g_{i-\varepsilon}(t)$. It may be represented by a superposition of functions $\phi_i(x,t;x_0)$ which are the densities of single diffusion processes started at $x_0 = i - 1 + \varepsilon$ or $x_0 = i - \varepsilon$ at time $t = 0$, with absorbing barriers at $x = i - 1$ and $x = i$. Function $\phi_i(x,t;x_0)$ has the same form as in Eq. (2):

$$f_i(x,t;\psi_i) = \phi(x,t;\psi_i) + \int_0^t g_{i-1+\varepsilon}(\tau)\phi(x, t - \tau; i - 1 + \varepsilon)d\tau$$

$$+ \int_0^t g_{i-\varepsilon}(\tau)\phi(x, t - \tau; i - \varepsilon)d\tau \qquad (4)$$

where the function ψ_i represents the initial conditions.

It is much easier to solve the system of all the above equations when they are inverted with the use of Laplace transform: all convolutions become in this case products of transforms, the Laplace transform of the function $\phi(x, t; x_0)$ is

$$\bar{\phi}(x, s; x_0) = \frac{\exp[\frac{\beta(x-x_0)}{\alpha}]}{A(s)} \sum_{n=-\infty}^{\infty} \left\{ \exp\left[-\frac{|x - x_0 - x'_n|}{\alpha} A(s)\right] \right.$$
$$\left. - \exp\left[-\frac{|x - x_0 - x''_n|}{\alpha} A(s)\right] \right\},$$

where $A(s) = \sqrt{\beta^2 + 2\alpha s}$.

The final solution $f_i(x, t; \psi_i)$ is obtained by numerical inversion of its Laplace transform $\bar{f}_i(x, s; \psi_i)$. In examples below we used Stehfest algorithm [9] where a function $f(t)$ is obtained from its transform $\bar{f}(s)$ for any fixed argument t as

$$f(t) = \frac{\ln 2}{2} \sum_{i=1}^{N} V_i \, \bar{f}\left(\frac{\ln 2}{t} i\right),$$

where

$$V_i = (-1)^{H/2+i} \times \sum_{k=\lfloor \frac{i+1}{2} \rfloor}^{\min(i, H/2)} \frac{k^{H/2+1}(2k)!}{(H/2 - k)!k!(k-1)!(i-k)!(2k-i)!},$$

and H is an even integer depending on a computer precision; we used $H = 14$.

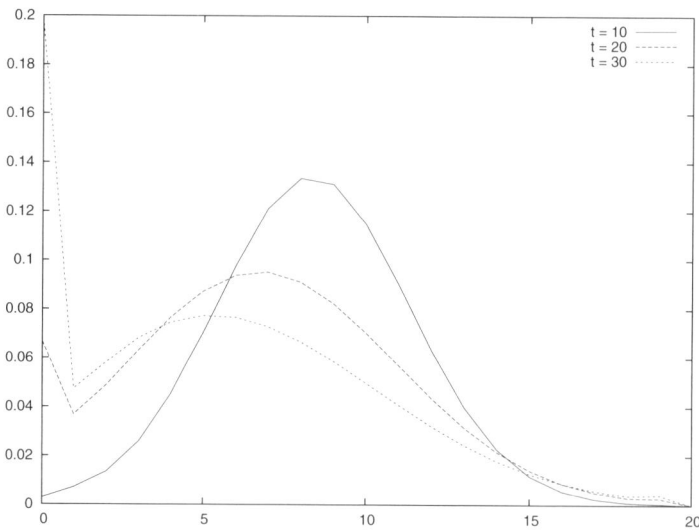

Fig. 5. The density function $f(x, t; x_0)$ of the distance to destination at time $t = 10, 20$ and 30; $\beta = -0.4$ and $\alpha = 0.54$; $x_0 = 10$

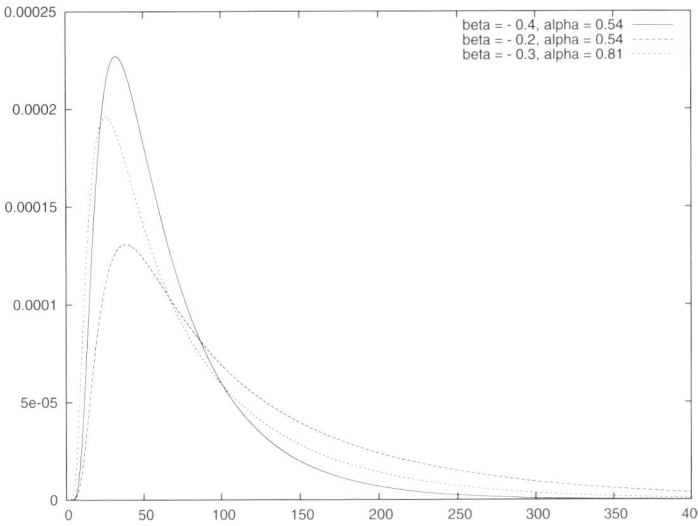

Fig. 6. The density function $\gamma_{x_0,0}(t)$ of the first passage time from x_0 to $x = 0$, i.e. the approximation of the transmission time density

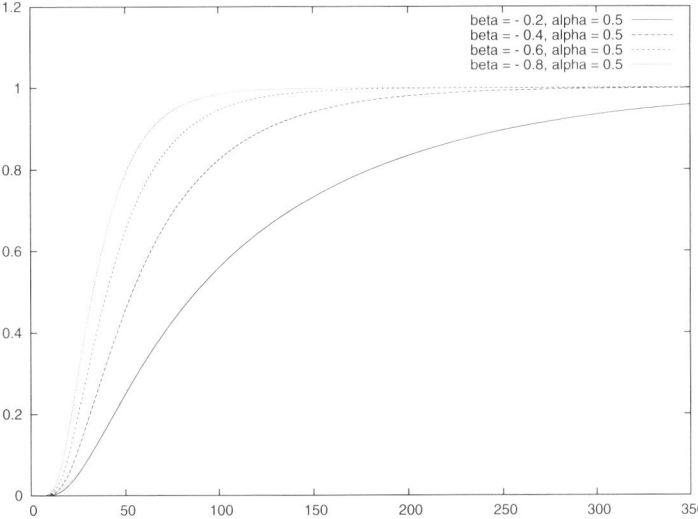

Fig. 7. Probability $p(0, t)$ if the starting point is $x_0 = 10$, diffusion interval $x \in [0, 20]$, $\alpha = 0.5$, and β is variable ($\beta = 0.2, 0.4, 0.6, 0.8$)

Some numerical results presented in Figs. 5 - 9 refer to the diffusion model with constant for the whole diffusion interval parameters. A model with parameters specific for each interval will be presented in the next section.

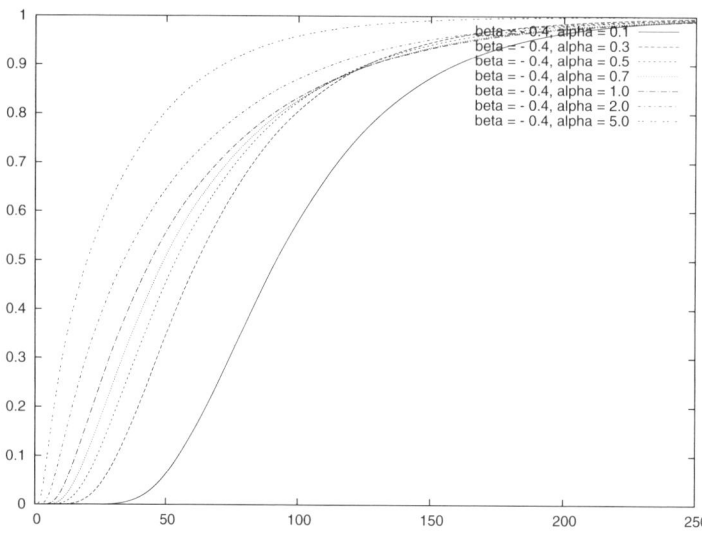

Fig. 8. Probability $p(0, t)$ if the starting point is $x_0 = 10$, diffusion interval $x \in [0, 20]$, $\beta = 0.4$, and α is variable ($\alpha = 0.1, 0.3, 0.5, 0.7, 1, 2, 5$)

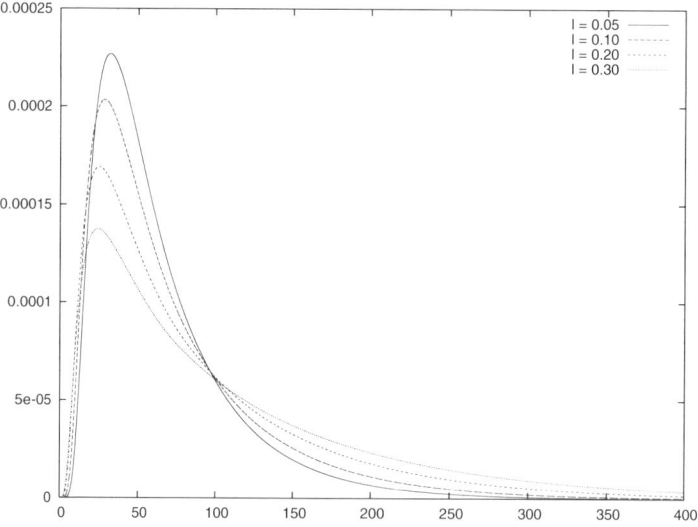

Fig. 9. The impact of loss ratio l ($l = 0.05, 0.1, 0.2, 0.3$) on the density function of transfer time at a network with $\beta = -0.4$ and $\alpha = 0.54$

Fig. 5 shows the density $f(x, t; x_0)$ of the remaining distance to complete the transfer calculated for time moments $t = 10, 20$ and 30 if $\beta = -0.4$ and $\alpha = 0.54$. The packet transmission is started at the distance $x_0 = 10$ from its destination.

Figure 6 displays the density $\gamma_{x_0,0}(t)$ of the first passage time from x_0 to $x = 0$, hence the approximation of the transmission time density. Some other exemplary curves of $p(0, t)$ for various parameters β and α are displayed in Figs. 7 and 8. As we might expect, larger α, i.e. larger variation of the movement, results in important tail of the transfer time distribution. The higher absolute value of the negative β shifts the maximum of the density towards zero and makes higher the probability that the transfer is finished before a specified time. Also the growth of α increases this probability.

Figure 9 presents the influence of the loss rate l ($l = 0.05, 0.1, 0.2, 0.3$) on the density of transmission time for fixed parameters $\beta = -0.4$ and $\alpha = 0.54$. We see an important sensibility of transmission time on the loss rate which is not surprising as each loss implicates a supplementary transmission between nodes.

6   Validation of the Model

As mentioned previously, parameters of the diffusion equation are determined by the mean and variance of the approximated discrete process. If π_{-1} is the probability to advance (to go to a node nearer to the sink by one hop), π_0 is the probability to stay at the same distance from the sink, and π_{+1} denotes the probability to go to a node more distant by one hop from the sink, then

$$\beta = \frac{\pi_{-1} \times (-1) + \pi_0 \times (0) + \pi_{+1} \times (+1)}{1 \text{ time unit}}$$

and

$$\alpha = \frac{\pi_{-1} \times (-1)^2 + \pi_0 \times 0^2 + \pi_{+1}(+1)^2}{1 \text{ time unit}} - \beta^2.$$

For example, in displayed numerical examples, the correspondence between these sets of parameters is as follows

$$\beta = -0.4, \quad \alpha = 0.54 \longrightarrow (\pi_{-1}, \pi_0, \pi_{+1}) = (0.55, 0.30, 0.15),$$
$$\beta = -0.2, \quad \alpha = 0.54 \longrightarrow (\pi_{-1}, \pi_0, \pi_{+1}) = (0.40, 0.40, 0, 20),$$
$$\beta = -0.3, \quad \alpha = 0.81 \longrightarrow (\pi_{-1}, \pi_0, \pi_{+1}) = (0.60, 0.10, 0.30).$$

Figs 10, 11 compare the results of diffusion approximation with simulation. The simulation model forms a line of nodes, transmission is started at a node N and a packet is forwarded to the node 1 (sink) with fixed probabilities $\pi_{+1}, \pi_0, \pi_{-1}$. The simulation results given for a certain time t are the means of 100 000 independent transmissions of a single packet. Fig. 10 refers to the probability density function of the first passage time $\gamma_{x_0,0}(t)$, hence to the approximation of packet travel time density, Fig. 11 refers to the probability $p(0, t) = \int_0^t \gamma_{x_0,0}(t)dt$ that the transmission is finished before time t, hence to the distribution function of the packet travel time.

Now we consider a two-dimensional network. Denote by S a sensor network characterized by a graph $G = <V, E>$ where V is the set of nodes and $E \subseteq V \times V$ is indicating the couples of nodes that can directly communicate. We group the nodes of S according to their minimal distance to the sink N_S. Define the subset $S_0 = \{N_S\}$. The subset $S_1 = \{N_i \in V/(N_S, N_i) \in E\}$ contains the nodes that are at the distance of one hop from the sink.

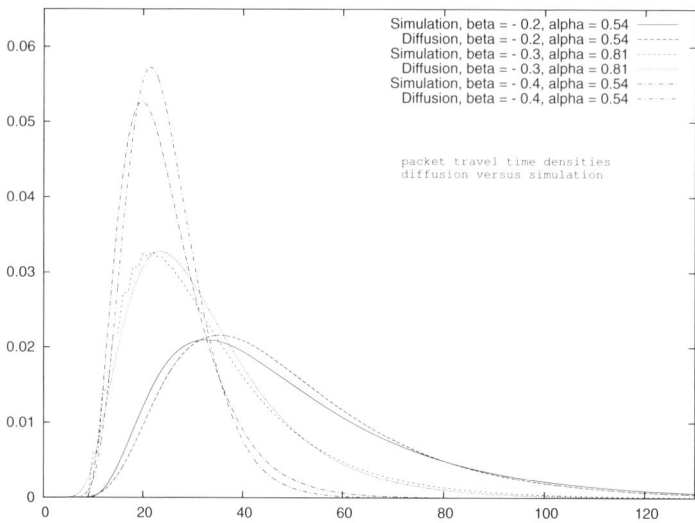

Fig. 10. Probability density function of packet travel time, comparison diffusion versus simulation, $x_0 = 10$ ($\beta = -0.4$, $\alpha = 0.54$), ($\beta = -0.2$, $\alpha = 0.54$), ($\beta = -0.3$, $\alpha = 0.81$)

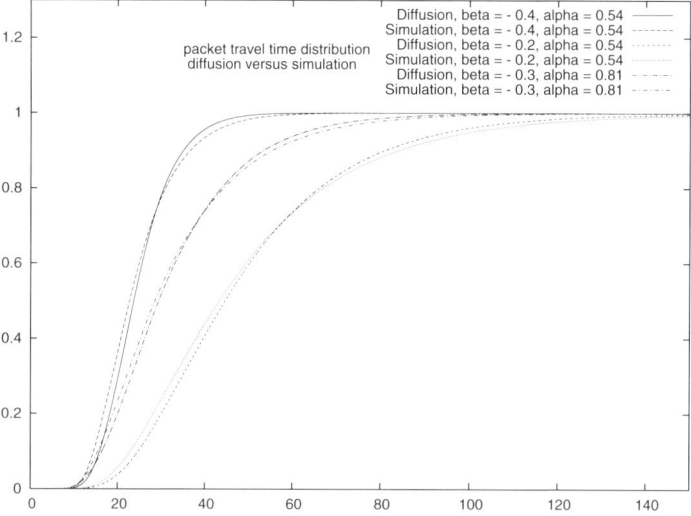

Fig. 11. Packet travel time distribution, comparison diffusion versus simulation, $x_0 = 10$, ($\beta = -0.4$, $\alpha = 0.54$), ($\beta = -0.2$, $\alpha = 0.54$), ($\beta = -0.3$, $\alpha = 0.81$)

Iteratively, $S_i = \{N_j \in V / \exists N_k \in S_{i-1}$ such that $(N_j, N_k) \in E\}$ includes all nodes having distance i to the sink. As the network is of finite dimension, the subset S_D, where D is the network diameter, contains the nodes most distant from the sink.

This partitioning leads to a particular graph where a node $N_i \in S_j$, $0 < j < D$, has neighbors only in S_j, S_{j-1} and S_{j+1}. All neighbors of N_S are in S_1 and those

Table 1. Mesh network example, parameters of the diffusion model

i	1	2	3	4	5	6	7
β_i	0.000000	−0.250000	−0.333333	−0.375000	−0.400000	−0.416667	−0.428571
α_i	1.000000	0.937500	0.888889	0.859375	0.840000	0.826389	0.816327
i	8	9	10	11	12	13	14
β_i	−0.437500	−0.444444	−0.450000	−0.454545	−0.458333	−0.461538	−0.464286
α_i	0.808594	0.802469	0.797500	0.793388	0.789931	0.786982	0.784439
i	15	16	17	18	19	20	21
β_i	−0.466667	−0.468750	−0.470588	−0.472222	−0.473684	−0.475000	−0.476190
α_i	0.782222	0.780273	0.778547	0.777006	0.775623	0.774375	0.773243
i	22	23	24	15	26		
β_i	0.477273	−0.478261	−0.479167	−0.480000	−0.480769		
α_i	0.772211	0.771267	0.770399	0.769600	0.768861		

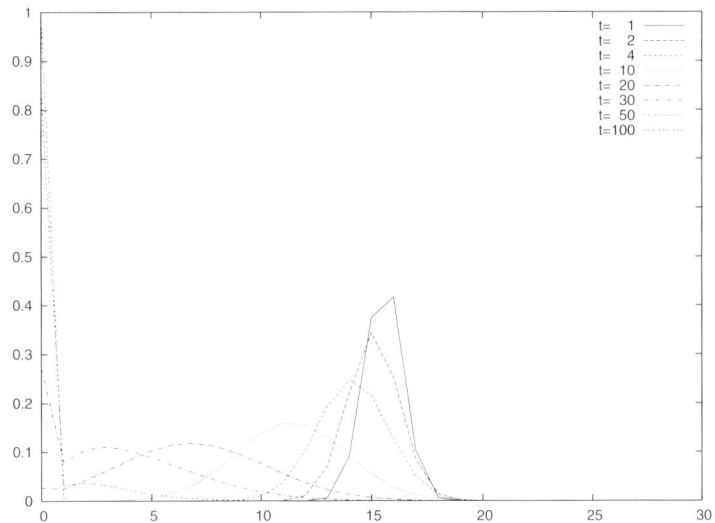

Fig. 12. Mesh network example, density of the diffusion process for chosen times

of S_D are in S_{D-1} or in S_D. Some protocols allow the nodes to know what is their distance to the sink as well as that distance for each one of neighbors [1]. A survey on sensor networks]. Hence, this partitioning is easy to implement and the routing from a node N_E to N_S is done by choosing at each time one of neighbors being closer to the sink. In general, the neighbors of $N_i \in S_j$ may be divided into $S_-(N_i) \subset S_{j-1}$, $S_+(N_i) \subset S_{j+1}$ and $S_=(N_i) \subset S_j$. If we fix the node N_i, we may easily determine the routing probabilities of reducing, leaving the same or increasing the distance to the sink. Normally, a node in N_i will choose a neighbor in N_{i-1}.

However, to save energy, the nodes might be a part of time in sleeping mode and cannot receive packets. Suppose that the nodes start their sleeping period following a certain schedule and inform about it their neighbors. If all the nodes of $S_-(N_i)$ fall

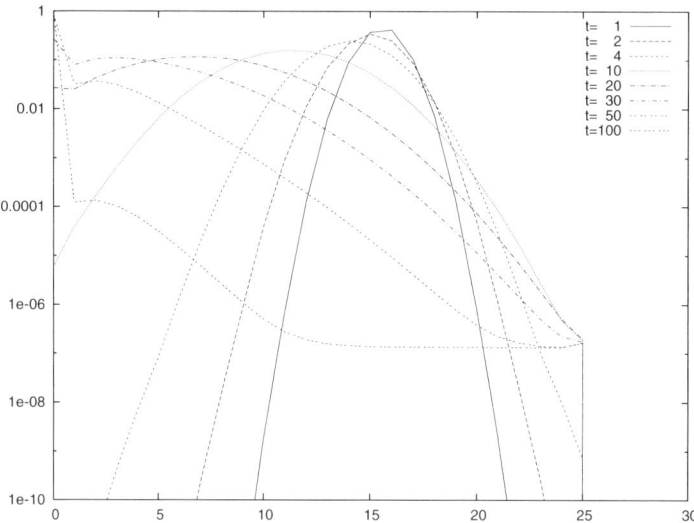

Fig. 13. Mesh network example, density of the diffusion process for chosen times, logarithmic scale

asleep, the packet should be sent elsewhere. Let ν be the fraction of time a node stays asleep. Hence, the probability $\pi_-(N_i)$ of decreasing the distance during the packet routing at N_i is $\pi_-(N_i) = 1 - \nu^{|S_-(N_i)|}$, where $|S_-(N_i)|$ is the cardinality of the set of neighbors of N_i being nearer to sink than N_i. Similarily, the probability of routing to a node at the same distance is $\pi_0(N_i) = \nu^{|S_-(N_i)|} \times (1 - \nu^{|S_=(N_i)|}$, and $\pi_+(N_i) = 1 - \pi_-(N_i) - \pi_0(N_i)$.

We are interested in routing probabilities for a set of nodes being at a distance j, hence $\pi_-(S_j), \pi_0(S_j), \pi_+(S_j)$, for $j = 1, \ldots, D$. They depend on the network topology and paths between the transmitter and the sink. In a general case this could be a complex task but for some regular topologies it is not difficult.

Let us consider as an example a two-dimensional mesh topology and, without any loss of generality, assume the unitary distance among neighboring nodes. Suppose also that the range of transmission does not allow to communicate neighbors in diagonal. Let (x_E, y_E) be the coordinates of the starting point and (x_S, y_S) the ones of the sink. Two cases are possible for intermediate nodes N_i placed at (x_i, y_i):

(i) $x_i \neq x_E$ and $y_i \neq y_E$: in this case 2 neighbors are in $S_-(N_i)$ and two others in $S_+(N_i)$. Hence, $\pi_-(N_i) = (1 - \nu^2)$ and $\pi_+(N_i) = \nu^2$;
(ii) $x_i = x_E$ or $y_i = y_E$: only one of four neighbors decreases the distance and three others neighbors increase it, so $\pi_-(N_i) = 1 - \nu$ and $\pi_+(N_i) = \nu$; Note that this formulation does not allow to have all neighbors in the sleeping mode and suppose implicitly that there is at least one active neighbor in $S_+(N_i)$ if all nodes in $S_-(Ni)$ are sleeping. However, to admit the case of all neigbors may fall asleep in the same time is also possible and conduct to a slightly different model.

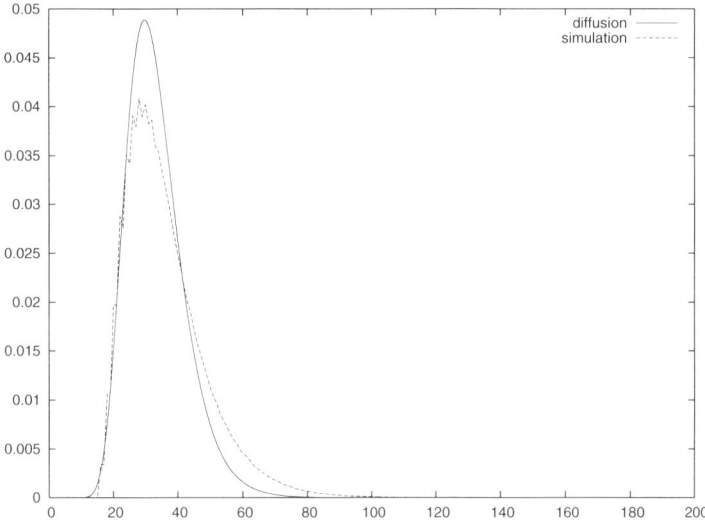

Fig. 14. Mesh network example, density of the transport time, diffusion and simulation results

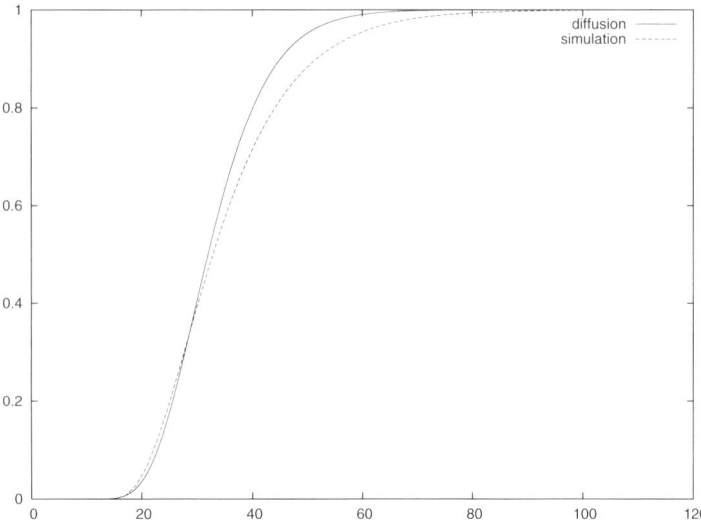

Fig. 15. Mesh network example, distribution of transport time, diffusion and simulation results

For the estimation of parameters $\pi_-(S_j)$ and $\pi_+(S_j)$ for the subsets of nodes being at the distance j from the sink, observe that in each S_j there are always four nodes having one coordinate same as the sink and that the cardinality of S_j is $|S_j| = 4j$. If we neglect the effects at the borders of the mesh, we obtain $\pi_-(S_j)$ and $\pi_+(S_j)$, $j > 0$:

$$\pi_-(S_j) = (1-\nu+(j-1)(1-\nu^2))/j, \qquad \pi_+(S_J) = (\nu+(j-1)\nu^2)/j, \qquad \pi_0(S_j)=0$$

Thus the parameters β and α of the diffusion model are:

$$\beta_j = \pi_+(S_j) - \pi_-(S_j), \qquad \alpha_j = 1 - \beta_j^2$$

Numerical example. Consider a mesh network having 26×26 nodes. The initial node is at $(x_E, y_E) = (18, 18)$, the sink is situated at $(x_S, y_S) = (10, 10)$, and $\nu = 1/2$. Observe that for each S_j we have different $\pi_-(S_j)$ and $\pi_+(S_j)$, hence also the diffusion process parameters are varying. Their values are given in Table 1. The densities of the diffusion process are displayed in Figs. 12, 13. The latter figure displays the results in logarithmic scale to show very small probabilities obtained via diffusion model. Then the density and the distribution of first passage time is compared with the same functions of the transport time obtained with simulation model, see Figs. 14, 15.

7   Conclusions

Owing to the introduction of the transient state analysis, the presented model captures more parameters (time-dependent and heterogeneous transmission, the presence of specific to each hop losses) of a sensor network transmission time than the already existing models of this type. It gives also more detailed results: the density function of a packet travel time instead of its mean value. Numerical results prove that the model is operational and the errors introduced by the method are reasonable.

Acknowledgements

This research was financed by Polish Ministry of Science and Higher Education project no. N517 025 31/2997 and supported by European Network of Excellence EuroFGI (Future Generation Internet).

References

1. Akyıldız, I.F., Su, W., Sankarasubramaniam, Y., Çayırcı, E.: A survey on sensor networks. IEEE Commun. Mag. 40(8), 102–114 (2002)
2. Cox, R.P., Miller, H.D.: The Theory of Stochastic Processes. Chapman and Hall, London (1965)
3. Gelenbe, E.: On Approximate Computer Systems Models. J. ACM 22(2) (1975)
4. Gelenbe, E.: Diffusion approximations: waiting times and batch arrival. Acta Informatica 12, 285–303 (1979)
5. Gelenbe, E.: Travel delay in a large wireless ad hoc network. In: 2nd Workshop on Spatial Stochastic Models of Wireless Networks (SPASWIN), Boston, April 7 (2006)
6. Gelenbe, E.: A Diffusion Model for Packet Travel time in a Random Multihop Medium. ACM Trans. on Sensor Networks 3(2), Article 10 (June 2007)
7. Heinzelman, W.R., Candrakasan, A., Balakrishnan, H.: Energy-efficient Communication Protocols for Wireless Microsensor Networks. In: Proc. Hawaiian International Conf. on Systems Science (January 2000)

8. Royer, E., Toh, C.: A review of current routing protocols for ad-hoc mobile wireless networks. IEEE Personal Commun. 6(2), 46–55 (1999)
9. Stehfest, H.: Algorithm 368: Numeric inversion of Laplace transform. Comm. of ACM 13(1), 47–49 (1970)
10. Zorzi, M., Rao, R.: Geographic random forwarding (GERAF) for ad hoc and sensor networks: Multihop performance. IEEE Trans. Mobile Comput. 2(4), 337–348 (2003)

Minimizing Interference in TDMA MAC Protocols for WSN Operating in Shadow-Fading Channels

Mário Macedo[1,2], Mário Nunes[1,3], and António Grilo[1,3]

[1] INESC, Rua Alves Redol, Nº 9,
1000-029 Lisboa, Portugal
[2] FCT/UNL,
2829-516 Caparica, Portugal
[3] IST/UTL, Av. Rovisco Pais,
1096 Lisboa, Portugal
`mmm@fct.unl.pt`
`{mario.nunes,antonio.grilo}@inesc.pt`

Abstract. This paper presents an analysis of the impact of shadow fading on the performance of TDMA slot allocation, with latter being measured in terms of the resultant in-network interference versus spatial reutilization. Simulations are presented which demonstrate that for Wireless Sensor Networks operating in shadow fading channels, protocols that feature TDMA slot allocation based on channel probing, e.g. LEMMA, lead to a significantly lower interferences when compared with protocols based on the n-hop neighborhood criterion. An extension of the basic LEMMA protocol is also presented, which leads to almost zero interferences.

Keywords: Wireless Sensor Networks, TDMA MAC protocols, Slot Allocation, Interference Avoidance, Shadow Fading Channel.

1 Introduction

Wireless Sensor Networks (WSN) has been a field of increasing research effort in the last years, and several dozens of MAC protocols have been designed for operation on WSNs. The major part of these protocols can be classified as contention-based, or TDMA-based, or even as combinations of these main approaches. Contention-based protocols present some specific sources of inefficiency such as idle listening, collisions, message overhearing, and control packets overhead. TDMA protocols, on the other hand, are well suited to avoid these problems, but require tight synchronization, and often require complex and sometimes message intensive slot assignment algorithms to guarantee collision and interference free slot schedules.

In this paper, we focus on the slot assignment procedures that are used to avoid collisions and interferences in the communication slots. A large number of TDMA protocols, such as [1], [2], [3], and [4], among others, use an n-hop (usually 2-hops) neighborhood criterion that can be described as following: a node can allocate a slot that is not previously occupied by its n-hop neighbors, hoping that this procedure is

L. Cerdà-Alabern (Ed.): Wireless and Mobility 2008, LNCS 5122, pp. 26–36, 2008.

sufficient to avoid collisions and interferences. However, this approach does not work well in some types of networks, namely for those irregular in shape, as we noted in a previous work [5]. Moreover, this procedure has the disadvantage of being message intensive.

Some other TDMA MAC protocols depart from the n-hop neighborhood criterion, but rely on different node transmission powers in order to avoid or identify and reduce possible interferences on the slots. For instance, RID (Radio Interference Detection in Wireless Sensor Networks, see [6]) aims to eliminate possible interference in the slots, by transmitting two detections packets: one transmitted at higher power (HD packet), followed by another transmitted at normal power (ND). With the transmitting node's identification given in the HD packet, and by measuring the power level at the reception of the ND packet, the receiver nodes can therefore estimate the possible interfering nodes. However, these procedures are very complex and computationally intensive.

Motivated by these observations, we designed the LEMMA (Latency-Energy Minimization Medium Access, see [5]) protocol, a new TDMA-based MAC protocol that uses cascading time-slot allocation to minimize latency while still achieving a very low duty-cycle. Although cascading TDMA slot allocation is not new, LEMMA bases slot assignment decisions on the interference really experienced by the WSN nodes through the carrier-sense mechanism. Its slot allocation procedure consists in probing the slots for any activity, both in the father node side, and the child node side, and trying to allocate the slot through messaging exchanges between them, such that other competing pairs for the same slot lose contention and quit allocating the slot. In order to detect activity, and therefore the occupancy of the slots, and to implement the carrier-sense mechanism, the LEMMA protocol uses the Clear Channel Assessment (CCA) functionality that is common in most WSN motes radios, namely the CC2420 radio of the MicaZ mote [14]. Reference [5] presents a description of the LEMMA protocol procedures and messages, and therefore they are not presented again in this paper.

Moreover, the behavior of the n-hop protocols is not studied in the presence of radio irregularity patterns, which are common in real world scenarios. In this paper, we study by simulations, the behavior of the n-hop protocols, and of the LEMMA protocol, in the presence of radio irregularity.

2   Radio Irregularity Propagation Models

Several authors (c.f., [7] and [8]), argue that the unit radius model, that is commonly used, namely for MAC performance analysis, is simplistic and not accurate, and that the radiation pattern is not circular, but quite irregular in shape. These variations are due to phenomena like reflections, diffraction, and scattering.

A commonly used model that accounts for these effects, is the shadowing model referred in [7], which can be described by the following expression:

$$PL(d)[dB] = PL(d_0)[dB] + 10 \times n \times \log_{10}\left(\frac{d}{d_0}\right) + X_\sigma \tag{1}$$

Where $PL(d)$ is the path loss at distance d, $PL(d_0)$ is the path loss at the reference distance d_0, n is the path loss exponent (which values 2 for the free space model), and X_σ is a random variable, given in dB units, with a normal distribution of zero mean and standard deviation of σ. This last term adds irregularity to the radiation pattern, because it introduces a variation from the isotropic pattern of the two first terms of the second half of the expression. Higher standard deviations lead to higher irregularities in the radiation patterns. Several authors, such as [9], measured values of up to 4 dB, for the standard deviation of X_σ, in WSNs scenarios.

Another model that accounts for the anisotropic path losses is RIM (Radio Irregularity Model) presented in reference [8]. In the RIM model, the received signal strength can be modeled by expressions (2) through (4):

$$\mathrm{Re}\,cSignalStrength = SendPower - DOIAdjustedPathLoss + Fading \qquad (2)$$

Where the *DOIAdjustedPathLoss* (*DOI* stands for Degree of Irregularity) can be computed as following:

$$DOIAdjustedPathLoss = PathLoss \times K_i \qquad (3)$$

K_i is a random irregularity factor, valuing around 1, and i is an integer index variable ranging between 0 and 359, for each degree of the plane. The K_i factors are calculated by expression (4):

$$k_i = 1, \text{ if } i{=}0$$
$$Ki = k_{i-1} \pm Rand \times DOI, \text{ if } 0{<}i{<}360, \qquad (4)$$
$$\text{Where } \left| k_0 - k_{359} \right| \leq DOI \text{ must hold.}$$

In expression (4), *DOI* is an experimentally estimated parameter, which gives the degree of irregularity of a given transmitting node, in some scenario; *Rand* is a continuous random variable that follows a Weibull distribution; the signs $\pm$ are randomly set.

The authors relate experiences that showed the ability of the RIM model to generate radio patterns that are adequate to model the observed data. One property of this model is that it generates continuous patterns, as the authors say that they are the observed ones. In that paper values are also presented for the *DOI* parameter, for several MicaZ nodes, which vary between 0.015, and 0.03. Adequate Weibull distribution parameters are also given.

3 Simulation Model

This section describes how the behavior of the n-hop neighborhood slot allocation criterion, and of the LEMMA interference avoidance procedure, was assessed in the presence of radio irregularity. As many protocols usually use a 2-hop criterion, the simulations that are presented in this paper were also made for this particular case.

A 100-node square grid was setup in the simulator, with the sink node placed at the upper-right corner. A tree topology was assumed, with the sink being the root node, and each node communicating with a random neighbor, selected among those that were closest, and that offered more progress towards the sink. In this way, each node was allowed to choose as parent either the node that was closest to it in the West direction, or in the North direction. Routing of packets was therefore straightforward: each node sent the packets to its parent, till the sink node. However, in this paper, simulation of the routing protocol was not assessed. Only the individual communications held between pairs of parent and child nodes were tested for possible signal interferences coming from another pairs of parent and child nodes that used the same timeslot and that were due to irregular radio propagation.

In order to set the distance between the nodes, a result from reference [10] was used, which states that the RSSI (received signal strength), needed to have a high packet receiving rate (PRR), should be higher than –93 dBm, for the MicaZ motes. The minimum received power was thus set to –90 dBm, to allow a safety margin of 3 dB, and the transmitting power of the MicaZ motes was set to the maximum value, 0 dBm. For simplicity of our simulation model, we did not account for the effects of the transitional regions of the radio links, in terms of the relation between PRR and SNR. Links that fall in those regions presents lower PRR, and are well known to be unstable. In our model, we considered for simplicity a sharp communication transition.

From expression (1), but not considering the random term X_σ, the following expression can be derived:

$$d = d_0 \times 10^{\wedge} \left(\frac{PL(d) - PL(d_0)}{10 \times n} \right) \qquad (5)$$

The reference distance d_0 was 1 meter, for which we had a value of –60 dBm of received power, or a path loss of 60 dB, value that was provided by some previous experiences in our labs, and which agrees with the values that are referred by other authors. Then, with a path loss of 90 dB for $PL(d)$, expression (5) can be used to estimate the maximum communication range, d, which is here designated d_{comm}. For instance, for a path loss exponent of 2, expression (5) gives a value of 31.6 m for d_{comm}. The dimension of the sides of the grid squares was set to $0.85 \times d_{\text{comm}} / \sqrt{2}$, in order to let nodes located diagonally to communicate directly. The distances between the nodes were therefore set to be dependent on the path loss exponents, allowing that each node could have always the same number of 1-hop, and 2-hop neighbors, when no radio irregularity was considered. As the main focus was on studying the effect of different slot allocation criteria in the number of interferences that were due to radio irregularity, the effect of the density of the nodes in the number of interferences was not studied.

In order to estimate the possible interferences, we used a result mentioned in reference [11], which states that for having a high PRR, near 1, the signal-to-noise ratio (SNR) should be at least 9 dB. Therefore, a value of –99 dBm (i.e., -90 dBm for communication, plus –9 dBm for the signal-to-interference margin) could be set for the threshold of the signal strength, above which it can interfere with the normal communication that is established in the slot.

Expression (5) can also give the minimum distance ($d_{\text{not\_int}}$) at which another node should be, in order to not interfere to a communication that is established in a given slot, between two nodes that are at a distance of d_{tx}:

$$d_{\text{not\_int}} \geq d_{\text{tx}} \times 10^{\wedge}\left(\frac{9\text{dB}}{10 \times n}\right) \qquad (6)$$

For instance, for a path loss exponent n of 2, relation (6) yields $d_{\text{not\_int}} \geq 2.8 \times d_{\text{tx}}$. This result shows that a 2-hop criterion is enough to avoid interferences in a simple model that does not account for irregularity. For higher path loss exponents, the obtained relations are more favorable, because the signal strengths decay faster. However, this is not true for lower path loss exponents, as we will show later.

The first set of our simulations used the more conventional shadowing of expression (1). As radio irregularity was included in the model, the criterion that we used for the computation of the 2-hop neighborhood was not geometric-based, but rather communication-based, and it was defined as following: the 1-hop neighbors of a given node, are those nodes that can communicate directly with that node; the 2-hop neighbors of a given node, are the 1-hop neighbors, plus the nodes that can communicate directly with the 1-hop neighbors.

The simulations began to firstly allocating randomly the slots in a distributed breadth-first manner. The interference avoidance procedure of the 2-hop neighborhood slot allocation criterion consisted in not using the slots that were previously occupied by the 2-hop neighbors of the node, and by the 2-hop neighbors of the father node. For the LEMMA protocol the shadowing model was used along with the interference avoidance procedure. If a pair of nodes, of a father and a child, saw no signal level above the interference level of –99 dBm, when they tested the slot for occupancy, they considered the slot as not occupied. The shadowing model was used to estimate the signal strength at each side of the pair of the father and the child.

Finally, we ran a test to determine if, for all pairs of father and child, there were transmitted signals, coming from other nodes that used the same slot, with strengths such that the 9 dB signal-to-interference ratio was not respected, both in the father and the child side. In this procedure, the shadowing model was used again in order to estimate these possible interfering signal strengths.

In the simulations, a frame of 1 second of duration, and 68 slots were used. All simulations results were taken from a set of 500 simulation runs. By statistical analysis, it is easy to show that this number of simulation runs can lead to small estimation errors, as it is mentioned later in the paper.

4   Simulation Results and Discussion

Fig. 1 shows the average number of interferences that were obtained by the simulation software, for different path loss exponents, and different standard deviations of the log-normal model, and for the 2-hop criterion. And Fig. 2 shows the respective results for the LEMMA interference avoidance slot allocation procedure.

As it can be seen, when the irregularity of the transmission path losses is considered, both slot allocation methods fail to have all the slots free from interference, but the numbers of interferences obtained with the 2-hop criterion are several times higher than those obtained with the LEMMA method.

The number of interferences augments with lower path loss exponents, and with higher standard deviations. If we decrease the path loss exponent, the signal

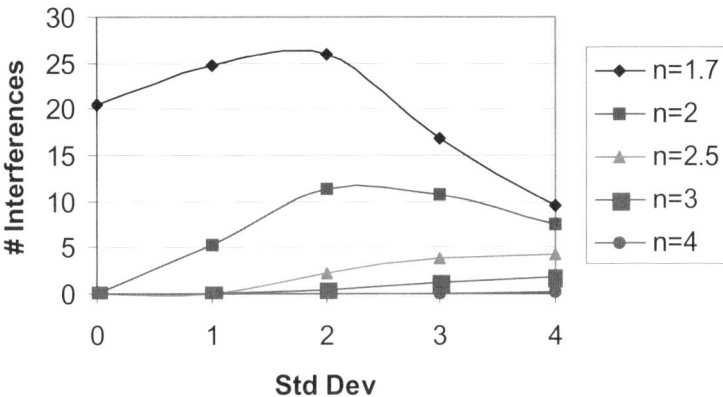

Fig. 1. Number of interferences of 2-hop neighborhood slot allocation criterion, when the shadowing model is added

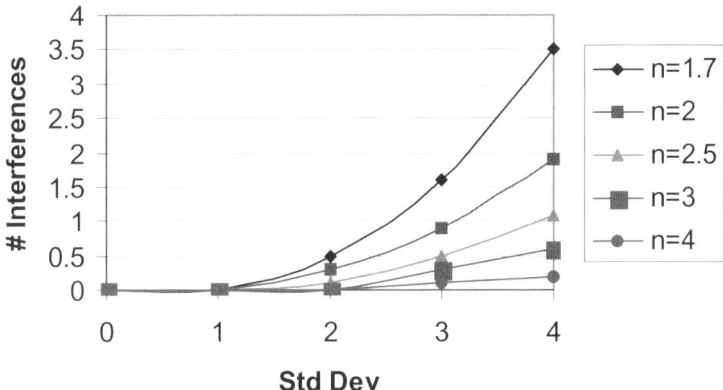

Fig. 2. Number of interferences of the LEMMA slot allocation procedure, when the shadowing model is added

attenuation is lower with the distance, and therefore the irregularity term of the shadowing model can be felt at higher distances.

Referring again to Fig. 1, it can be seen that the 2-hop criterion fails to avoid interferences, even when no irregularity is added (zero for standard deviation), when we have lower path loss exponents (in the case, valuing 1.7). This can be explained by solving expression (6), which yields $d_{not\_int} \geq 3.38 \times d_{tx}$ for a path loss exponent of 1.7. This relation shows that the 2-hop criterion is not enough to avoid interferences, because nodes situated at a distance of $\sqrt{3^2 + 1^2} \times d_{tx} = 3.16 \times d_{tx}$ can still interfere. However, if we lower the signal-to-interference ratio to 8 dB, for which we can expect a PRR around 95%, according to reference [11], the 2-hop criterion would also present no interferences, for the no radio irregularity scenario, because expression (6) would yield $d_{not\_int} \geq 2.96 \times d_{tx}$.

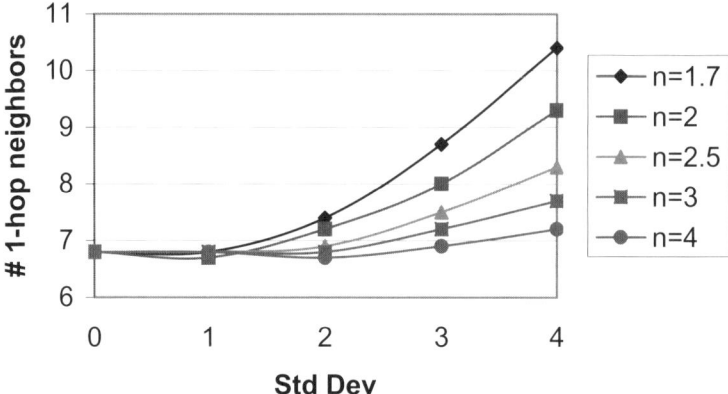

Fig. 3. Number of 1-hop neighbors obtained for different path loss exponents and irregularities

In order to highlight the relative accuracy of the simulation results presented here, it is worth to mention that, for instance, the value of the number of interferences that were obtained for the 2-hop criterion, for $n = 1.7$, and $\sigma = 2$, was equal to 26, for which it was calculated a 95% confidence interval of [25.46; 26.54].

The higher number of interferences that is obtained for the 2-hop criterion seems to be due to a geometric reason: avoiding slots that are used in a 2-hop communication neighborhood is not sufficient in some cases to avoid interferences, since the distance of interference is several times higher than the distance of communication, and not merely twice that value. Adding irregularity seems to worsen even more this effect. When irregularity is added, some transmitted signals can become stronger, or weaker. It is possible that the signal transmitted by one node is not strong enough to make it a 2-hop neighbor of another node, but still being strong enough to interfere with its reception.

Another observed phenomenon is that for lower path loss exponents, the number of interferences experienced by the 2-hop criterion begins to decrease for high values of σ. The higher number of 2-hop neighbors that are present when the irregularity increases explains this. This last phenomenon is also referred and predicted by mathematical analysis in reference [12], and Fig. 3, and 4 of our simulations corroborate it. But when the neighborhoods of a given node grow, the nodes that could interfere begin to be very far, and therefore have lower probability to do so.

Note, however, that this increase on the number of neighbors does not mean that adding irregularity to the radio propagation model leads necessarily to an increase on the global network connectivity. Simulations reported in a complementary study of reference [12] (see, [13]) show the opposite, i.e., that when the irregularity term X_{σ} rises, the connectivity of the networks decreases. This phenomenon is explained by the fact that bi-directional links become more asymmetric, and more links become unidirectional, therefore deteriorating the performance of the routing protocols. In the case of our simulations, we counted as neighbors the nodes that could communicate some information, such as *hello* packets, or slot occupancy information, in the case of the 2-hop criterion, but we did not require that these links were bi-directional. The

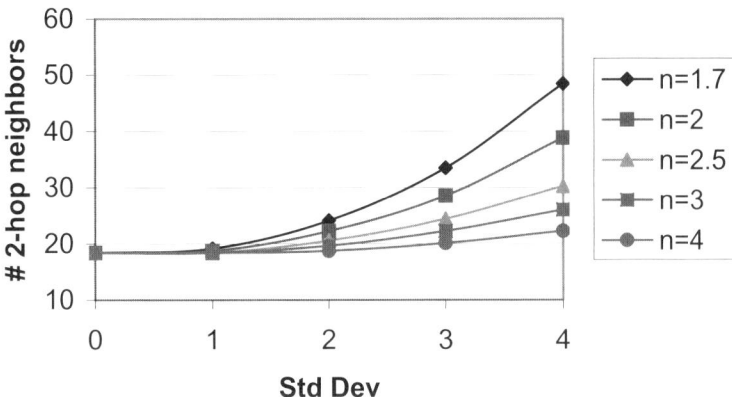

Fig. 4. Number of 2-hop neighbors obtained for different path loss exponents and irregularities

only links that we required to be bi-directional were those that were established between a pair of a father and a son in the logical routing tree of our simulations. These nodes were placed close enough to each other, to avoid that the bi-directionality of their respective links is significantly affected by increasing radio irregularity.

A possible objection to the model that was used in these simulations could be that of not using a spatial correlated log-normal shadowing model. But, firstly, the results reported in reference [8] seem to show that the radio propagation patterns of the motes have a small correlation distance, when compared with the respective distances between them. Secondly, we admit that, in most scenarios, the dimensions of the obstacles can be small when compared with the distances between the motes. This observations support our option of not using a more sophisticated spatial correlated shadowing model.

Another objection could be that of not using Rayleigh fading, or the more appropriate for a dominant line-of-sight, Rician fading [7]. But as these irregularity terms present faster spatial variations, and also occur around the mean value of the path loss of the shadowing fading model, we expected that these more accurate models would not affect significantly the simulation results presented here.

Returning back to the simulation results, it is also worth to note that higher radio irregularities are usually present for higher path loss exponents (see [12], for this observation). Therefore, the highest values of interferences that were obtained for the basic LEMMA protocol simulations, and shown in the upper-right side of Fig. 2 – which were present in the scenarios with higher values of σ, and with the lower path loss exponent values of 1.7, and 2 – correspond to situations that are probably rarer. This argument supports also the conclusion that the basic LEMMA protocol is much better than the 2-hop criterion in terms of number of resulting interferences.

We also assessed the behavior of the two slot allocation methods using the RIM radio irregularity model, instead of the log-normal shadowing model. Fig. 5 shows the results of the simulations that were obtained for the different path loss exponents. In these simulations, we used the distributions of the *DOI* that were presented in reference [8], for several MicaZ motes.

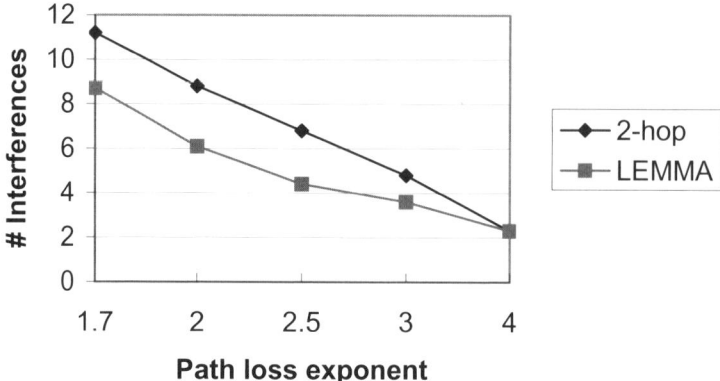

Fig. 5. Number of interferences obtained with RIM radio irregularity model for the 2-hop slot allocation criterion and for the LEMMA protocol, for different path loss exponents

Although LEMMA still shows superiority in comparison to the 2-hop criterion, this superiority seems to be less evident than that shown with the log-normal model. However, we have two objections to the RIM model: we feel that the RIM model is excessively dependent on a specific propagation scenario, because the irregularity term depends on field measurements of the *DOI* for each specific mote. On the contrary, the simulations made with log-normal shadowing model should cover a broader range of scenarios. Another somehow bizarre particularity of the RIM model is that of the irregularity term of expressions (2) through (4) being dependent on each particular transmitter, instead of being more related to the propagation environments. Therefore, we do not give so much credit to the results of this last set of simulations.

Regarding to the interferences that are present with the LEMMA protocol, they result from the asymmetry of the links: suppose that a pair of father and child had previously allocated a slot, and another pair of father and child achieve to allocate the same slot, because they do not feel it as occupied. Due to asymmetry of the transmissions, that arises when radio irregularity is added, this last pair can interfere with the communications of the former pair.

We are working on a solution to this problem. It consists in adding tones, to the basic LEMMA protocol allocation messages. These tones are transmitted at a higher power level than that of the protocol messages, and were proved to be able to overcome the asymmetry of the links, leading to almost zero interferences.

5   Conclusions and Future Work

This paper has presented a comparison between TDMA slot allocation based on LEMMA, and protocols that employ the *n*-hop neighborhood criterion. Simulations lead to the conclusion that in real environments, the LEMMA protocol will show a substantial advantage over the more commonly used 2-hop neighborhood criterion, in terms of number of interferences that persist after slot allocation. Interferences in the LEMMA protocol are due to asymmetric links, while interferences in the 2-hop criterion are also inherent to its geometric nature. The 2-hop neighborhood slot allocation criterion presents an intolerable number of interferences when a more realistic radio propagation model is considered, and when a realistic signal-to-interference ratio is used. This criterion also presents problems when more irregular shape networks are considered.

We also presented an extension to the basic LEMMA that achieves almost zero interferences.

For future work we intend to continue the study of the behavior of the extended LEMMA protocol in the presence of radio propagation irregularity. Another direction for future work is the design of slot allocation strategies and heuristics that minimize the latency of the communication from the nodes to the sinks.

Acknowledgments. This work was supported in part by the EuroFGI ("Design and Engineering of the Future Generation Internet"). Apart from this, the European Commission has no responsibility for the content of this paper.

The information in this document is provided as is and no guarantee or warranty is given that the information is fit for any particular purpose. The user thereof uses the information at its sole risk and liability.

References

1. Bao, L., Garcia-Luna-Aceves, J.J.: A New Approach to Channel Access Scheduling for Ad Hoc Networks. In: Proceedings of the Seventh Annual International Conference on Mobile Computing and Networking, ACM MobiCom 2001, Rome, Italy (July 2001)
2. Rajendran, V., Obraczka, K., Garcia-Luna-Aceves, J.J.: Energy-Efficient, Collision-Free Medium Access Control for Wireless Sensor Networks. In: Proceedings of the ACM SenSys 2003, Los Angeles, USA (2003)
3. Rajendran, V., Garcia-Luna-Aceves, J.J., Obraczka, K.: Energy-Efficient, Application-Aware Medium Access for Sensor Networks. In: Proceedings of 2nd IEEE International Conference on Mobile Ad-Hoc and Sensor Systems (IEEE MASS 2005), Washington, USA (November 2005)
4. van Hoesel, L., Havinga, P.: A Lightweight Medium Access Protocol (LMAC) for Wireless Sensor Networks. In: Proceedings of the First International Workshop on Networked Sensing Systems (INSS 2004), Tokyo, Japan (June 2004)
5. Nunes, M., Grilo, A., Macedo, M.: Interference-Free TDMA Slot Allocation in Wireless Sensor Networks. In: Proceedings of the 32nd IEEE Conference on Local Computer Networks (LCN 2007), Dublin, Ireland, October 2007, pp. 239–241 (2007)
6. Zhou, G., He, T., Stankovic, J.A., Abdlezaher, T.: RID: Radio Interference Detection in Wireless Sensor Networks. In: Proceedings of the 24th Annual Joint Conference of the IEEE Computer and Communications Societies (INFOCOM 2005), Miami, USA (March 2005)
7. Rappaport, T.: Wireless Communications: Principles and Practice, 2nd edn. Prentice Hall, Englewood Cliffs (2002)
8. Zhou, G., He, T., Krishnamurthy, S., Stankovic, J.A.: Models and Solutions for Radio Irregularity in Wireless Sensor Networks. ACM Transactions on Sensor Networks 2(2), 221–262 (2006)

9. Martinez-Sala, A., Molina-Garcia-Pardo, J.M., Egea-Lopez, E., Vales-Alonso, J., Juan-Llacer, L., Garcia-Haro, J.: An Accurate Radio Channel Model for Wireless Sensor Networks Simulation. Journal of Communications and Networks 7, Part 4, 401–407 (2005)
10. Srinivasan, K., Levis, P.: RSSI is Under Appreciated. In: Proceedings of the Third Workshop on Embedded Networked Sensors (EmNets 2006), May 2006, Harvard University, Cambridge (2006)
11. Lee, H., Cerpa, A., Levis, P.: Improving Wireless Simulation through Noise Modeling. In: Proceedings of the 6th International Conference on Information Processing in Sensor Networks (IPSN 2007), Cambridge, Massachusetts, USA (April 2007)
12. Bettstetter, C., Hartmann, C.: Connectivity of Wireless Multihop Networks in a Shadow Fading Environment. Wireless Networks 11(5), 571–579 (2005)
13. Stuedi, P., Chinellato, O., Alonso, G.: Connectivity in the Presence of Shadowing in 802.11 Ad Hoc Networks. In: Proceedings of the IEEE Wireless Communications and Networking Conference (WCNC 2005), New Orleans, LA, USA (March 2005)
14. "MICAz datasheet", Crossbow (date of last access, November 16, 2007), http://www.xbow.com/Products/productdetails.aspx?sid=164

Secure and Efficient Data Collection in Sensor Networks

Cristina Cano[1], Manel Guerrero[2], and Boris Bellalta[1]

[1] Universitat Pompeu Fabra (UPF)
Edifici Estació França, 08003-Barcelona, Spain
{cristina.cano,boris.bellalta}@upf.edu
[2] Universitat Politècnica de Catalunya (UPC)
Computer Architecture Department
Jordi Girona 1-3, E-08034 Barcelona, Spain
guerrero@ac.upc.edu

Abstract. Sensor networks are a very specific type of wireless networks where both security and performance issues need to be solved efficiently in order to avoid manipulations of the sensed data and at the same time minimize the battery energy consumption. This paper proposes an efficient way to perform data collection by grouping the sensors in aggregation zones, allowing the aggregators to process the sensed data inside the aggregation zone in order to minimize the amount of transmissions to the sink. Moreover, the paper provides a security mechanism based on hash chains to secure data transmissions in networks with low capability sensors and without the requirements of an instantaneous source authentication.

1 Introduction

Nowadays, the manufacture of inexpensive wireless sensor nodes powered by batteries opens a broad range of applications [1] (from environment observation to health applications, from home to big commercial applications, etc.). Nevertheless, the means by which those sensor networks collect the information they sense in such a way that it will be efficient and that the sensor nodes will not run out of battery too soon are still not in place. Routing protocols for general purpose wireless networks (like AODV [2] and OLSR [3]) cannot be used efficiently in wireless sensor networks due to the specific characteristics of their control traffic. Moreover, the use of security will be fundamental for some common sensor applications, that typically are implemented with computationally expensive cryptographic primitives.

Therefore, the design of security mechanisms for sensor networks has to take into account that sensors are, typically, nodes with very limited memory and computer power that, additionally, can lack of a node identifier. In this paper, we propose the utilization of *hash chains* as a way of providing delayed authentication but satisfying the low computation capabilities of the sensors without requiring any type of key pre-settings. In addition, the hash can be used to

L. Cerdà-Alabern (Ed.): Wireless and Mobility 2008, LNCS 5122, pp. 37–48, 2008.

identify the sensor nodes in the network. Hash chains have being used as an efficient way to obtain authentication in several approaches that tried to secure routing protocols. For example, in [4], [5] and [6] they use them in order to provide delayed key disclosure. While, in [7], hash chains are used to create one-time signatures that can be verified immediately. The main drawback of all the above approaches is that all of them require clock synchronization. SAODV uses hash chains to authenticate hop counts [8,9]. In SEAD [10] (by Hu, Johnson and Perrig) hash chains are also used in combination with DSDV-SQ [11] in a very similar way (this time to authenticate both hop counts and sequence numbers). At every given time each node has its own hash chain. The hash chain is divided into segments, elements in a segment are used to secure hop counts in a similar way as it is done in SAODV. The size of the hash chain is determined when it is generated. After using all the elements of the hash chain a new one must be computed. SEAD can be, in theory, used with any suitable authentication and key distribution scheme. But finding such a scheme is not straightforward.

Additionally, in this paper, a new simple tree discovery and routing through the tree algorithm is proposed. A solution to improve the efficiency in data transmission is the use of aggregation techniques (i.e., join several data packets in a single one in order to reduce the unnecessary overhead transmitted). This can result in lower battery consumption and an increment of capacity (in messages transmitted by the network with the same transmission resources). Krishnamachari et al. analyze, in [12], from a theoretical point of view the benefits and drawbacks of doing data aggregation in wireless sensor networks. Nevertheless, they do not show a way of performing that data aggregation. Kalpakis et al. present, in [13], an algorithm to solve the data collection problem for wireless sensor networks. Nevertheless, the algorithm they provide is polynomial-time. In SIA [14], Przydatek et al. consider information aggregation in sensor networks but they assume that there are some nodes that perform aggregation and other nodes that do not, and leave how to decide which nodes are aggregators and which are not out of the scope of their paper.

Finally, we conclude this paper with an evaluation of the new network layer with security and data aggregation in order to asses what are the costs of using security in terms of performance and the benefits of using aggregation or data fusion.

2   Building a Tree with Aggregation Zones

In a typical sensor network, there is a 'sink' (a node that collects the information that the other nodes sense) and the sensor nodes. Thus, the streams of information have the structure of a tree that has the sink node as its root, and which is probably one of the most efficient structure for data collection and aggregation for sensor networks. The challenges are, however, the construction or discovery of the tree, the grade of security provided and how efficiently the different streams can be merged when converge in the same path.

2.1   The Discovery of the Tree

In a sensor network with no or slow node mobility, the discovery of the 'collector tree' will only need to be done after nodes deployment and when the sink notices that it stops receiving an important part of the sensor nodes' reports (which might be due to the dead of one or many sensor nodes or due to a change in the link connectivity). To perform a tree discovery, the sink broadcasts a Tree Discovery ('TD') message. Every node that receives it marks the sender as its father [1] in the tree and unicasts a 'YAMF' (You Are My Father) reply back to the node. Therefore, once the tree discovery is over, every node knows which node is its father and which nodes are its children. Figure 1.(a) shows how TD messages get propagated, and Figure 1.(b) shows how YAMF messages travel back.

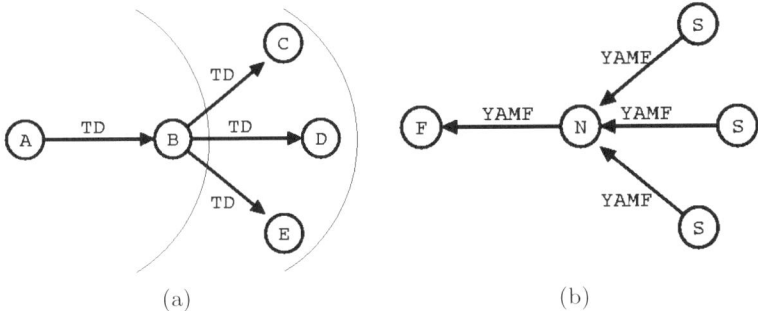

Fig. 1. (a) Propagation of the TD message, (b) Propagation of the YAMF message

The 'TD' message will include information of whether the sensor nodes have to report with what they sense periodically, or only as a reaction to certain event. Additionally, it can include information regarding the role of an specific group of sensors (for example, if there are powered nodes which must take always the role of aggregators).

2.2   Adopting a Node

In the case that a node detects that it has lost its link connectivity with its tree father or a new node joins the network, it can broadcast an 'ILMF' (I've Lost My Father) message to its neighbors. A node that receives an 'ILMF' message will forward it towards the sink in the same manner as it would do with a message containing sensed data. When the sink receives the 'ILMF' message or messages it will decide whether to trigger a new tree discovery or not. With this the sink can consider to trigger a new 'TD'.

---

[1] About node identifiers, we consider that sensors neither have a pre-shared identifier nor negotiates it from the sink. Thus, nodes generate a random value, with enough length to be statistically unique, which will be used as node identifer.

A node that issues an 'ILMF' message because it has realized that its father is not reachable anymore or it is joining the network might set an 'Adoption' flag in the message. This flag indicates that the node is willing to accept a neighbor as its new father. A neighbor receiving an 'ILMF' message will offer to adopt the node and it will issue an 'IWTA' (I'm Willing To Adopt you) message. The orphan node will chose among the nodes who sent him the 'IWTA' (typically it will choose the first one) and will notify it by sending to its future father an 'YMNF' (You're My New Father) message. In the case that no node is willing to adopt the orphan node, the node will issue another 'ILMF', but this time without the 'Adoption' flag set. Notice that, the process to adopt a node, to make it very simple, is independent of the aggregation process.

2.3   Aggregation

Intermediate nodes are capable of aggregating data (see Figure 2, where an sketch of the network topology is shown). These nodes can be standard sensor nodes that, for example, with a probability p change to an aggregation role during a certain period of time (this information can be disseminated in the 'TD' message). However, due to the requirements to authenticate the data received before aggregate it, an special consideration for them is required. These nodes aggregate the data received in a fixed duration time period (D_{agg}) by performing some kind of data fusion (for instance computing the mean of the received data). Moreover, they must perform data authentication in a similar way as it would be done at the sink.

The use of aggregation will reduce the transmitted information, and thus, the power consumption is also reduced. Additionally, with respect to the MAC

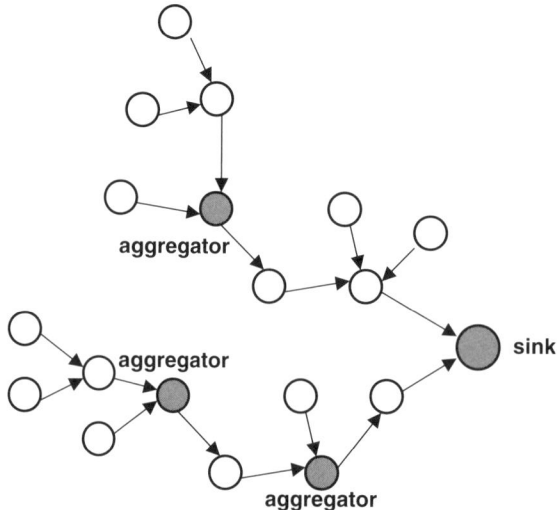

Fig. 2. Sketch of the tree with aggregation nodes scenario

protocol, the overall network performance will be improved as the use of aggregation will decrease - in average - the number of packets transmitted to the channel for each node. For the specific case of random access MAC protocol (ex. CSMA), which is the most commonly used in wireless sensor networks, this will reduce the probability that two transmitted packets from different sensors collide, improving the efficiency of the channel bandwidth use.

3   Solving the Data Authentication

A malicious node must only be able to give wrong data about its sensor data and to decide whether it forwards or not the sensor data it is supposed to forward towards the sink. Both cases are detectable by the aggregation nodes or the sink. Thus, there are two security services which will be guaranteed:

- **Integrity:** The sink needs to be able to verify that the information that it is being reported by a sensor node has not been altered by a forwarding node.
- **Source authentication:** In addition, the sink needs to be able to verify that a node providing the data is the one it claims to be.
- The two last security services combined build **data authentication**.

Availability is outside of the scope of this paper. Although of course it would be desirable, it does not seem to be feasible to prevent denial-of-service attacks in a network that uses wireless technology (where an attacker can focus on the physical layer without bothering to study the routing protocol). Notice that it is assumed that **Confidentiality** is not required.

It is important to look at some possible attacks to the presented protocol and to other protocols designed to be used in sensor networks.

- **Multiple Personality:** In this attack a node pretends to be **n** nodes. Every time it is supposed to provide an information it does for all of its multiple personalities. If **n** is not too small in comparison to the total number of sensors in the network, then the perception of the situation by the sink will be very misled.
- **Man-in-the-middle:** In this attack the node has multiple personality of as many nodes as sons, grandsons, etc., he has. When it receives the sensed data by them it modifies the data at its will and uses its other personalities to forward the modified sensed data.

3.1   Using Secret Keys and/or Public Key Cryptography

Those two previous attacks must be taken into account when studying how to implement source authentication. It could be argued that the only way for a sink to detect such attacks, would be to know (in a way) all the sensor nodes or the use of a secret key known between each of the nodes and the sink.

Nevertheless, if the requirement of sharing a secret key between each of the sensor nodes and the sink before network deployment is not a feasible one, an alternative method could be used: The sink would have a key pair and its public key would be known by all the sensor nodes. After network establishment the sensor network would collect the secret key of each of the sensor nodes encrypted with the public key of the sink in the same manner sensed information is collected. The main problems with this approach are that attacks of the kind of 'Multiple personality' are not avoided and that public key cryptography is computationally expensive.

3.2   Using Hash Chains

If it is acceptable that the sensed data that a node sends is not authenticated immediately, but with the next data transmission, hash chains could be used to obtain delayed authentication. With this technique, the identity of a node gets marked by its TOP-HASH, which in fact, is also used as the node identifier.

To construct a hash chain one must choose a hash function and generate a 'seed' (typically a random number that must be kept secret) of the same length than the hash numbers that the hash function generates. The seed will be the first 'link' (element or hash) of the hash chain. Then, one calculates each of the links by hashing the previous link. So, the second link is the result of hashing the seed; the third link is the result of hashing the second link; etc. Until the chain has the required number of links (length). We will call the last link of the hash chain 'TOP-HASH'.

This approach will not solve the problem with attacks of the kind of 'Multiple personality' but requires no pre-sharing of any keys, data or whatsoever. On the other hand, it will limit the number of sensed data messages that sensor nodes can send in an authenticated way with the same hash chain, re-computing it (if allowed) when all the hashes has been used. Nevertheless, the number of hashes will, arguably, be higher than the messages that it will be sent before battery deployment for most scenarios. For example, a pollution metering with 1 measures each hour, during 1 year would send less than 10 K measures, which is expected to be shorter than the hash chain considered. Anyway, the hash chain could be designed specifically for each situation and recomputed when all its values are used if its required.

Each node generates a seed for the hash chain, from which it is able to compute all the links of the chain from the seed to the TOP-HASH (see Figure 3).

The number of hashes or 'links' with its correspondent position that will be stored (if a sensor hasn't enough memory to store all the chain) depend on two

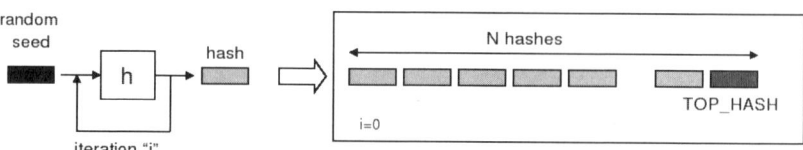

Fig. 3. Hash Chain Computation

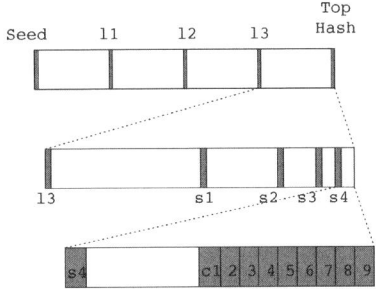

Fig. 4. Example of stored links of a hash chain: Seed, l1..l3, s1..s4, c1..c9, and Top Hash

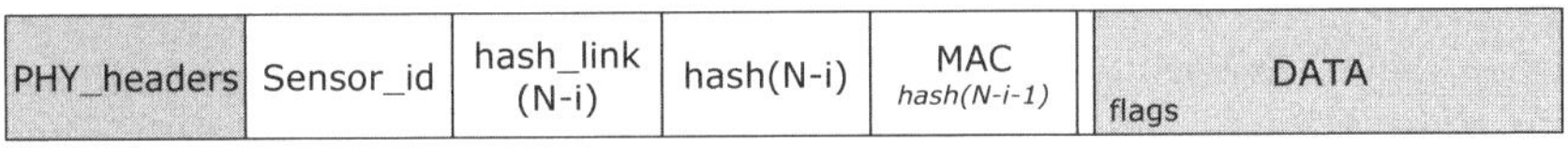

PHY_headers	Sensor_id	hash_link (N-i)	hash(N-i)	MAC hash(N-i-1)	flags	DATA

Fig. 5. Packet Format (transmission i)

things: the available memory for storage and the maximum number of hashes that a node must calculate. Otherwise, only a group of links will be stored. In figure 4 there is an example of how this would work. Let's assume that the maximum number of hashes that it is desired that a node must calculate to obtain the last not used link of the chain (therefore the next element of the chain to be used) is n. Then, a node will store the seed (which is the 0th element of the chain), and the nth, $2n^{th}$, $3n^{th}$, etcetera, until the TOP-HASH ($l1$, $l2$ and $l3$ in the figure). If the node has the possibility to store more hash links it will store as many links that halve the last segment as possible ($s1$, $s2$, $s3$ and $s4$ in the figure) and so on.

A node will calculate the next hash link to be issued from the last stored hash link instead of starting from the seed. Every time the node computes the last link of the chain from the last stored $'s'$ link it will store the as many of the last non-stored links of the chains as c links. After all the $'s'$ hash links have been used, the node will calculate and store the $'s'$ links of the new last $'l'$ segment.

Each time a sensor nodes issues the i data message, it precedes the data that it has sensed by a node identifier, this node identifier can be the TOP-HASH since (when long enough) is going to be statistically unique. After the node identifier and the data, it will include a Message Authentication Code (MAC) of the information using the link $N - i - 1$ of the chain and revealing the link $N - i$ of the chain and the flags field is used to indicate if the packet contains data aggregated or not (the packet format is shown in Figure 5).

Then, each node that receives it will receive the current information and it will be able to authenticate the information sent in the previous message. Notice that, if some messages are lost, the sink can still perform delayed verification because it knows how many times the new hash value has to be hashed.

3.3   Considerations Using Aggregators

About how to merge the security requirements and aggregation at intermediary nodes, there are several considerations:

1. An aggregator node only will start to aggregate the received data after it has received the second transmission of an specific sensor. The first message of this specific sensor received by the aggregator is just forwarded to the sink without processing it (the aggregator only records the required hash-related fields in order to authenticate the next message received from that sensor) (a). The second message received from the specific sensor will be aggregated as it has been used to authenticate a previous message and it is expected that it will be authenticated also with next transmission. However, in order to authenticate the previous message by the sink, which requires the hash-based fields from this second message, the aggregator builds an extra packet with these fields and sends them to the sink (these fields could be concatenated to the same packet in which the data is aggregated) (b). Finally, all next packets are aggregated and authenticated at the aggregator (c). The process is depicted in Figure 6.

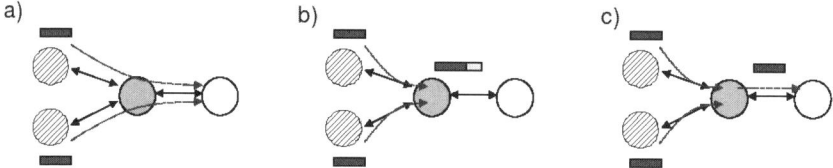

Fig. 6. Aggregation using security

2. Aggregators use its own hashes to authenticate the aggregated data. Probably, higher security requirements should be considered for those nodes.
3. Notice that this mechanism is only useful if the aggregation process or data fusion (e.g. to compute the average of a set of values, to concatenate measures without process them, etc.) allows it.
4. Aggregators only aggregate messages which have not been aggregated before. It means that packets which have been already aggregated will be only forwarded by the other aggregators in the path to the sink.

4   Performance Results

To evaluate the protocol previously described a network simulator has been developed using the COST (Component Oriented Simulation Toolkit) package [15]. A network formed by N sensor nodes distributed over a plain area has been simulated, each node generates λ packets of a fixed length L_{data} per second on average. It has been considered that the resulting fusion of all data received in the interval could be reduced to fit into a single packet of length L_{data}. The benefits of using aggregation in the intermediary nodes have been compared with the case in which all packets are forwarded to the sink. Moreover, the secure solution proposed has been compared with the non-secure case, the main difference between them is the length of data packets, in the secure solution the packet length is composed of five different fields (due to the use of delayed authentication with hash chains), then:

- Always - Sensor identifier (which can be the TOP-HASH) of 128 bits, which is the length of the $MD5$ hash.
- Only with security - The Message Authentication Code of the data (128 bits long).
- Only with security - Identifier of the link in the hash chain (4 bytes).
- Always - Flags (4 bits).
- Only with security - The hash link corresponding to the previous data (128 bits long).
- Always - Payload $L_{data} = 36$ Bytes.

The channel capacity between two nodes is in average equal to $C = 10\ Kbps$ over the wireless link and ideal channel conditions have been considered. Each sensor node generates and sends data at a rate equal to $\lambda = 1/10\ packets/s$. When a node is configured to aggregate packets, it waits for a certain amount of time $D_{agg} = 2.5$ s and sends only one packet containing the result obtained from the fusion function applied to the data received. On average, intermediary nodes are responsible of aggregating data of 8 sensors.

The results obtained from the simulations are shown in Figures 7, 8 and 9 with 95% confidence intervals. In Figure 7 the total traffic sent by the aggregation nodes is plotted. It could be seen that by using security the total load of the network increases but it could be reduced by using aggregation techniques. In

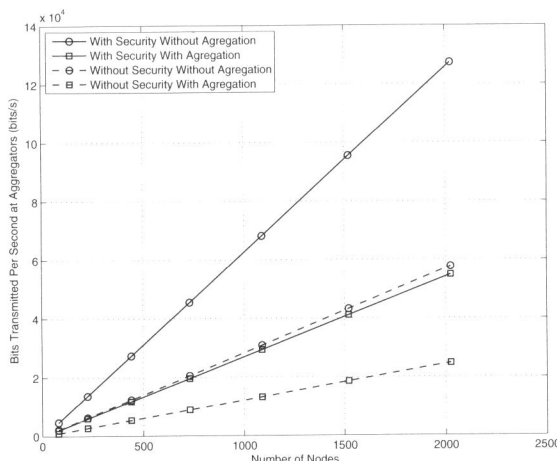

Fig. 7. Impact of security and aggregation on the information transmitted

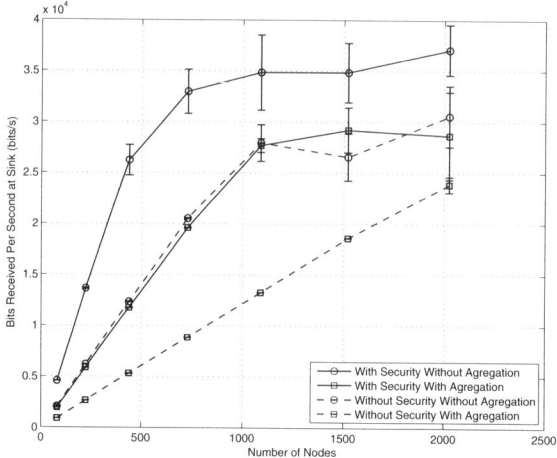

Fig. 8. Impact of security and aggregation on the information received

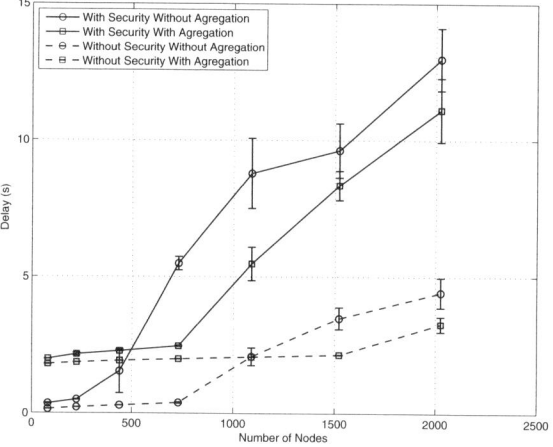

Fig. 9. Impact of security and aggregation on the packet delay

this particular case, the load of using security and aggregation is similar to the load obtained without security and without aggregation. Notice that when the network is not congested, the variance between the different simulations is quite small, which results in very small confidence intervals for those points.

Figure 8 shows the total traffic received (transmitted - lost) at sink. The saturation point of the network could be derived from this figure, notice that without aggregation the network saturates with approximately 400 nodes (with security) and 700 nodes (without security), but using aggregation the number of nodes supported increases to 700 using security and 1500 without security. Higher aggregation delays would allow to obtain a higher number of nodes in unsaturated conditions.

Figure 9 plots the average end-to-end delay for the scheme proposed with and without secure functions and aggregation. Notice how the delay increases noticeably when the saturation point is reached. The prize to be paid by using aggregation is an extra delay caused by the aggregation delay (D_{agg}). It is important to see that this delay is only higher when the saturation point is not reached, otherwise the aggregation delay is smaller than the delay obtained without using aggregation for the same number of nodes.

When adding security, the network load increases in both cases leading to a reduction in the network size. Therefore, the cost of using security has to be considered in terms of a reduction of the network size and/or the frequency of data transmission by the sensors.

5   Concluding Remarks

This work introduces a set of preliminary ideas to be applied in securing low-capability sensor networks. It shows an efficient way to perform data collection in sensor networks while providing different security mechanisms to secure data transmission depending on the needs of the sensor network and the capability of the nodes.

The network can be secured by using shared secrets between the sink and the sensor nodes that might be shared before or after network deployment or, when shared secrets between sink and sensor nodes before deployment is a too strong requirement and sensor nodes do not have the computational capacity to calculate signatures, by the use of hash chains.

Network efficiency (less battery consumption, higher network size, etc.) is improved by using an asynchronous data aggregation mechanism which is shown as a valid mechanism for non-real time sensor networks as it reduces the extra bits due to unnecessary overheads.

Finally, summarizing the main results / assumptions done:

– Very-low capability sensors.
– Simple protocols to guarantee the network connectivity and the aggregation functionality.
– No confidentially, only delayed-authentication is provided through the use of hash-chains.
– When using aggregation, no tracing of the source at the sink.
– Significant benefits when using data aggregation in terms of network size.

References

1. Akyildiz, I., Su, W., Sankarasubramaniam, Y., Cayirci, E.: A survey on sensor networks. IEEE Commun. Mag. 40(8), 102–114 (2002)
2. Perkins, C.E., Belding-Royer, E.M., Das, S.R.: Ad hoc on-demand distance vector (AODV) routing. Internet Request for Comments RFC 3561 (November 2003)

3. Clausen, T., Jacquet, P. (eds.) Adjih, C., Laouiti, A., Minet, P., Muhlethaler, P., Qayyum, A., Viennot, L.: Optimized link state routing protocol (olsr). RFC 3626, Network Working Group (October 2003)
4. Hauser, R., Przygienda, A., Tsudik, G.: Reducing the cost of security in link state routing. In: Symposium on Network and Distributed Systems Security (NDSS 1997), San Diego, California, February 1997, pp. 93–99. Internet Society (1997)
5. Cheung, S.: An efficient message authentication scheme for link state routing. In: 13th Annual Computer Security Applications Conference, pp. 90–98 (1997)
6. Perrig, A., Szewczyk, R., Wen, V., Culler, D.E., Tygar, J.D.: SPINS: security protocols for sensor netowrks. In: Proceedings of the 7th Annual International Conference on Mobile Computing and Networking, pp. 189–199 (2001)
7. Zhang, K.: Efficient protocols for signing routing messages. In: Proceedings of the Symposium on Network and Distributed Systems Security (NDSS 1998) (July 2001)
8. Asokan, N.: Presentation at an informal workshop on mobile and ad hoc networking security, EPFL, Lausanne (December 2001)
9. Zapata, M.G., Asokan, N.: Securing Ad hoc Routing Protocols. In: Proceedings of the 2002 ACM Workshop on Wireless Security (WiSe 2002), September 2002, pp. 1–10 (2002)
10. Hu, Y.C., Johnson, D., Perrig, A.: SEAD: Secure efficient distance vector routing for mobile wireless ad hoc networks. In: 4th IEEE Workshop on Mobile Computing Systems and Applications (WMCSA 2002), June 2002, pp. 3–13 (2002)
11. Broch, J., Maltz, D.A., Johnson, D.B., Hu, Y.C., Jetcheva, J.: A performance comparison of multi-hop wireless ad hoc network routing protocols. In: Proceedings of the 4th Annual International Conference on Mobile Computing and Networking, pp. 85–97 (1998)
12. Krishnamachari, B., Estrin, D., Wicker, S.B.: The impact of data aggregation in wireless sensor networks. In: ICDCSW 2002: Proceedings of the 22nd International Conference on Distributed Computing Systems, Washington, DC, USA, pp. 575–578. IEEE Computer Society, Los Alamitos (2002)
13. Kalpakis, K., Dasgupta, K., Namjoshi, P.: Efficient algorithms for maximum lifetime data gathering and aggregation in wireless sensor networks. Comput. Networks 42(6), 697–716 (2003)
14. Przydatek, B., Song, D., Perrig, A.: SIA: Secure information aggregation in sensor networks. In: ACM SenSys 2003 (November 2003)
15. Chen, G., Szymanski, B.K.: Cost: A component-oriented discrete event simulator. In: Proceedings of the 2002 Winter Simulation Conference, pp. 776–782 (2002)

Image Recognition Traffic Patterns for Wireless Multimedia Sensor Networks

Ruken Zilan[1], José M. Barceló-Ordinas[1], and Bülent Tavli[2]

[1] Universitat Politècnica de Catalunya (UPC)
Computer Architecture Department
Jordi Girona 1-3, E-08034 Barcelona, Spain
{rzilan,joseb}@ac.upc.edu
[2] TOBB Univ. of Economics and Technology
Computer Engineering Department
Ankara, Turkey
btavli@etu.edu.tr

Abstract. The objective of this work is to identify some of the traffic characteristics of Wireless Multimedia Sensor Networks (WMSN). Applications such as video surveillance sensor networks make use of new paradigms related with computer vision and image processing techniques. These sensors do not send whole video sequences to the wireless sensor network, but objects of interest detected by the camera. In order to able to design appropriate networking protocols, a better understanding of the traffic characteristics of these multimedia sensors is needed. In this work[1], we analyze the traffic differences between cameras that send whole coded images and those that first process and recognize objects of interest using Object Recognition techniques.

Keywords: Wireless Multimedia Sensor Networks, Object and Image Recognition, Traffic Patterns, Imagers, Image Compression, Video Coding.

1 Introduction

Wireless Multimedia Sensor Networks (WMSN) are gaining research interest due to the availability of low-cost cameras and CMOS image sensors, also due to the broad application requirements. Applications of such networks can be listed as: Multimedia surveillance sensor networks, advance health care delivery, personal locator services, traffic avoidance, enforcement and control systems, etc, [1].

WMSN may include the transmission of *snapshots* in which event triggered observations are obtained and *multimedia streaming* in which long multimedia data is sent requiring high data rates and low-power consumption transmission techniques. For example, we can imagine a surveillance application in which several cameras are deployed to control habitat monitoring. When a motion

---

[1] This work has been supported by Spanish Ministry of Science and Technology under grant TSI2007-66869-C02-01 and EuroFGI Network of Excellence.

L. Cerdà-Alabern (Ed.): Wireless and Mobility 2008, LNCS 5122, pp. 49–59, 2008.

sensor detects an animal, the camera sensor takes images and send the data to a sink using a multihop network. In order to minimize the power consumption and network lifetime, it would be interesting to first detect whether the image captures a phenomena of interest (e.g. a particular animal) and second send the minimum data to represent that object. Furthermore, it would be interesting to describe the phenomena from multiple views and on multiple resolutions.

In Wireless Sensor Networks, the traffic sent into the network consists of a few bytes of data. Most of the works consider that sensors react to events or to queries. As an example of typical parameters, in Directed Diffusion [2], the authors consider that each source generates two events per second and each event was modeled as a 64-byte packet while interests were modeled as 36-byte packets at a periodic rate of one interest every 5 seconds with a interest duration of 15 seconds. Another example is T-MAC [3], in which the authors use as a data model based on periodic packets of 50 bytes every second to the sink, packets of 30 bytes every 4 seconds in the neighborhood of the event (and event is produced every 10 seconds) and local packets every 20 seconds. As can be seen, in general traffic in sensor networks is modeled as periodic sources.

In Wireless Multimedia Sensor Networks traffic sources produce *snapshots* or *multimedia streaming*. In any case, if we take a camera and produce frames at a rate that may range between 5 and 30 fps (frame per second) the traffic produced will not be periodic. For instance, it is well known that video streams (e.g. MPEG coded) exhibit heavy-tailed probability distributions and autocorrelation functions with a mixture of Short Range Dependence (SRD) and Long Range Dependence (LRD). This kind of traffic in a sensor network may rapidly consume the sensor batteries and also will fill the buffers of the sensors. Object Recognition is a scientific discipline that focuses on obtaining specific information from images. This research area includes scene reconstruction, event detection, tracking, object recognition, etc. Given most of the applications present in WMSN, we believe that Object Recognition is a discipline that has to be taken into account in the design of most multimedia sensor networks. There are applications that instead of sending the whole set of frames, send a stream of bytes representing part of the scene, such as the edge of an object of interest in the scene (e.g. a person, a car, etc). That means the sensor will produce traffic based on the output of software that manipulates the frames taken from the scene. Many works related to traffic characterization on different network architectures may be found in the literature. O. Rose et al [4], studied the impact of MPEG1 video traffic in ATM networks with detailed statistical analysis. F. Fitzek et.al. [5], also presents statistical results from MPEG4 and H.263 encoded video streams for wire-line and wireless networks. Related to WMSN, C.F. Chiasserini et.al. [6], investigate the possible trade-offs between energy consumption and image quality. The authors mention the high complexity in terms of time and power that may result in performing motion estimation when coding frames. After that, the authors study JPEG performance in still images in which only intra-frame (Discrete Cosine Transform mainly) is performed. Power consumption in video sensor technologies are also studied in several video platforms proposed. Examples are Panoptes,

Meerkats and Cyclops [7], [8] and [9]. Meerkats [8], shows that energy dissipation on computation is no longer negligible and is comparable to the energy dissipation on communications.Meerkats utilizes and object detect based on background subtraction that simply detects motion taken between two snapshots at different short time lags. Cyclops [9], performs object recognition using background subtraction on images taken periodically. AER [10], takes another approach. AER (Address Event Representation) outputs only a few features of interest of the visual scene by detecting intensity differences (motion) information.

However, there is no works able to identify traffic characteristics that these sensor devices may produce. Even more, Meerkats chose DSR (Dynamic Source Routing) as a routing protocol for tests. Although DSR is not a suitable protocol for sensor networks, we think that in order to design appropriate sensor network protocols for video sensing, it is necessary to identify the traffic that these devices produce. Therefore, the objective of this work is to identify some of the traffic characteristics that may be found in Wireless Multimedia Sensor Networks (WMSN) that make use of new paradigms related with computer vision and image processing techniques. These sensors do not send whole video sequences to the wireless sensor network, but objects of interest detected by the camera. In order to, may be in further studies, able to design appropriate networking protocols, a better understanding of the traffic characteristics of these multimedia sensors is needed. In this work, we analyze the traffic differences between cameras that send whole coding images and those that first process and recognize objects of interest using Object Recognition techniques.

The paper is organized as follows, in section 2, Traffic Generation of Framework is explained under three subsections which are General Framework, Video Surveillance in WSN and Experimental Study. In section 3, Statistical Analysis are discussed. Finally in section 4, the paper is concluded.

2   Traffic Generation Framework

2.1   Object Recognition Basics

There is a wide application area for object recognition fields like industry (quality control etc.), medical image possessing, military applications and in the explorations of field and autonomous processing vehicles (cars, robots, etc.). Mainly, identification of an object and determination of its parameters are called Object Recognition. Although there are huge amount of techniques for object recognition depending on the applications, general object recognition methods can be categorized as in Figure 1.

There are typical functions that can be found in many object recognition techniques: *Image acquisition* (either 2D image, 3D volume or an image sequence), *pre-processing* (noise reduction, contrast enhancement and scale-space representation), *detection/segmentation* (one or multiple image regions), *feature extraction* (lines, edges and ridges, or more complex features such as texture, shape or motion related requirements), *high level processing*. Further information on object recognition basics can be found in [11].

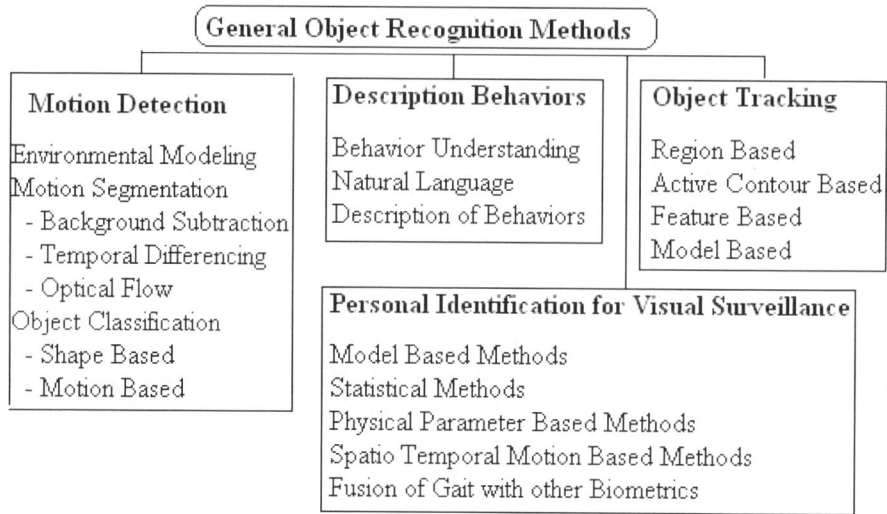

Fig. 1. Generation Object Recognition Methods

2.2 General Framework

The experimental set-up, see Figure 2, consists of a camera that takes video at a configurable rate of M fps (frames/s). Here, we consider two cases: frames f_1,

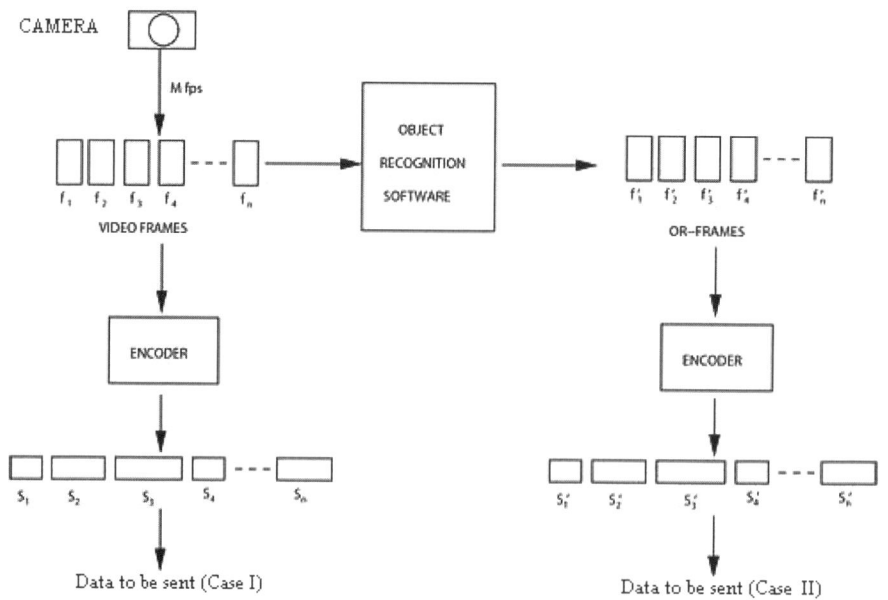

Fig. 2. Traffic Generation Framework

$f_2,..., f_n$ may be encoded and be sent to the network (Case I). After encoding, frames have a size (in bytes) S_1, $S_2,..., S_n$ that depends on the chosen coding technique. This is the data that should have to be sent to the network if the camera does not implement any detection technique. In the second case frames f_1, $f_2,..., f_n$ are fed to one of the OR tools. This output consists of a set of non-encoded OR-frames $f_1^{'}$, $f_2^{'},...., f_m^{'}$. In general, and depending on what does the Object Recognition (OR) tool, the number of OR-frames may be different from the original set (m $\neq$ s). For instance, some tools represent changes in light intensity with metadata. On the reception side the tools are able to represent detected objects from these metadata packets received. Other tools, subtract the background and outputs frames with the envelope of the object detected. Further compression may be obtained using any coding technique. Therefore, after encoding, OR-frames have a size (in bytes) $S_1^{'}$, $S_2^{'},...., S_m^{'}$.

2.3 Experimental Framework

In this study the chosen OR method is *Edge Detection*. It is one of the main steps in feature extraction, detection and segmentation. Edges could be considered as a boundary between two dissimilar regions in an image. Computation of edges are fairly cheap and recognition of an object is easy since it provides strong visual clues however edges can be affected by the noise in an image. Although there are several different methods to perform edge detection, generally they can be categorized under two subtitles; Gradient and Laplacian. While the first one detects the edges by looking for the maximum and minimum in the first derivative of the image, the second one searches for zero crossings in the second derivative of the image to find edges. For this study, *Sobel Edge Detection method* which can be categorized as gradient, is chosen. We have used a software called Video OCX [12] to capture live video as a gradient method and perform image processing. Although Video OCX and its tools are not designed for sensor networks, they are practical tools for image processing that allows capturing live video, displaying and saving AVI files. It also helps for applying image processing functions like edge detection based on Sobel filtering, or motion detection of images.

AER (Address Event Representation), developed by YALE University [10], is an address event sensor that extracts and outputs only a few features of interest from the visual scene. These sensors are sensitive to light and motion so that only pixels that include the brightest one generate the events first and more periodically. Thus, only pixels which experience a high enough difference in the light intensity generate the events. ENALAB at YALE University provides AER emulator for research purposes. AER emulator takes an image from the camera, after some low-level feature detection it produces a frame indicating the importance of the detected feature with the pixel value. After that, an algorithm converts each pixel's magnitude into frequency coding, by connecting it to the other pixels' own streams to produce a stream of address-events. AER uses the maximum event rate value to produce as many as events per pixels. Since AER is designed for sensor networks, we have used AER Emulator to produce traffic

Fig. 3. Original, Edge Detected and AER snapshots

representative of sensor imagers. Figure 3 presents a snapshot of the original sequence, edge detected sequence, and AER sequence.

After Object Recognition, AER processed frames are not optimally encoded in terms of spatial or temporal redundancies. Since, at the moment, no specific encoders are designed for sensor networks, we have used MPEG4. MPEG4 may not be suitable for sensor networks in terms of computation. In Wireless Sensor Networks, the transmission/reception module consumes more power than computation of data. In Wireless Multimedia Sensor Networks, computation of data (e.g. encoding) may be higher than in data-centric sensors. We have not found studies on suitable encoder for WMSN. Nevertheless MPEG encoding requires high computation (e.g. Discrete Cosine Transform, etc), we think that an encoder which includes spatial redundancy compression techniques may reduce the amount of data to be sent in a sensor network. Applying temporal redundancy compression is more difficult due to the need of higher computation capabilities and the need of buffering past/future frames (P and B frames).

The MPEG (Motion Picture Experts group) coding algorithm was developed mainly for the storage of compressed video on digital storage media. These standards are based on coding technologies was developed mainly for digital video compression for storage on digital media. The MPEG video compression algorithm uses two techniques [13]: block-based motion compensation for the reduction of the temporal redundancy and transform domain-(DCT) based on compression for the reduction of spatial redundancy. Motion compensated techniques are applied with both causal (pure predictive coding) and non-causal predictors (interpolative coding). The remaining signal (prediction error) is further compressed with spatial redundancy reduction (DCT). To encode frames, three modes can be used; intraframe (I), predictive (P) and interpolative (B). An I-frame is encoded as a single image, with no reference to any past or future frames. A P-frame is encoded relative to the past references frame. A reference frame is a P- or I-frame. A B-frame is encoded relative to the past reference frame, the future reference frame, or both frames (I or P).

3   Experimental Study

To study the traffic generated by the OR tools we characterize the frame sizes. For this purpose, we analyze three different sets of data: A live captured-video file, a captured and features detected by Edge Detection video file and a video file which is progressed with AER emulator. The scenario can be considered as a surveillance environment processed by the AER emulator in which a person walks in a room in front of the camera. The video sequence lasts around one minute. The experiment emulates a sensor camera that detects an object of interest (a person) and sends the captured images to a sink. The objective is to analyze the different kind of traffic this sensor could generate before sending the data to the sink.

The uncompressed 16-bit IYUV video is captured at different frame rates in the QCIF format (176x144 pxl). When using Edge Detection, files are captured in gray scale. Captured files, first encoded by MPEG4 encoder (IMToo Mpeg Encoder) and then frame sizes are extracted by the usage of MPEG-4 Parser [14], and statistics are taken from these trace files.

Table 1 shows the data sizes generated before and after encoding. At the moment of this work, the capturing software can only work with live video. Therefore, the input of the three methods are different even though they are of similar sequence and time and can not be compared directly. However, the results still are interesting, since they show the different traffic patterns produced by the three methods. In Table 1, we see that the results from different Frame Rates behave in a similar way. When we observe the 25 fps capture, we can see that the Original data to be sent without any compression is 89873 KB and using MPEG4 compression reduces to 4592 KB. Using Edge Detection the data to be sent without compression reduces to 21139 KB, due to the fact that Edge Detection is coding in gray scale using 2 bits/pixel. Further compression using an encoder such as MPEG4 reduces the amount of data to 4585 KB. Although the compression ratio of the original file is higher than the other, the amount of data to be sent is approximately the same. Note that we can not directly compare both outputs since they correspond to different input files (all around 1 minute) and we only consider the magnitude of the results as a first approximation.

However, traffic characteristics, see Table 2, are quite different. Original frames have a higher Peak to Mean Ratio, generally an indication of higher burstiness.

Table 1. Data sizes and Compression rates

Method	Frame Rate	Numb. Capt. Frames	YUV (KB)	MPEG4 (KB)	Compres. Rate
Original	25 fps	1164	89873	4592	19,57
EDGE	25 fps	1272	21139	4585	4,61
Original	20 fps	1272	98266	5012	19,61
EDGE	20 fps	1056	17774	4288	4,15
Original	15 fps	1176	91210	4661	4,38
EDGE	15 fps	804	13721	3133	4,38

Table 2. Data Frame Statistics

Method	Frame Rate	Mean Size (B)	St. Dev.	Coeff. Var.	Peak	Peak/Mean	Min.
Original	25 fps	3833	2561,6	66,82	14529	3,79	1307
EDGE	25 fps	3574	3042,2	85,1	8009	2,24	67
Original	20 fps	3818	2508,1	65,69	14027	3,67	1420
EDGE	20 fps	3978	3399	85,44	8820	2,22	69
Original	15 fps	3823	2585,6	67,63	16093	4,21	1391
EDGE	15 fps	3775	3240	85,83	8788	2,33	70

Table 3. I-Frame and P-Frame Statistics (25 fps)

Method	Type of Frames	Num. Frames	Mean Size (B)	St. Dev.	Coeff. Var	Peak	Peak/Mean	Min.
Original	I	97	11976	1191	9,94	14529	1,21	9865
Original	P	1067	3093	671	21,70	5949	1,92	1307
EDGE	I	106	6050	457	7,55	7149	1,18	5046
EDGE	P	1166	3349	3077	91,87	8009	2,39	67

Peak frames get to the order of 14000-15000 Bytes while Edge detection have peak frames of no more than 9000 Bytes. Separating, see Table 3, I-frames from P-frames helps to see the different compressions obtained in the traces.

After Object Recognition, VideoOCX saves captured video in AVI format. AER emulator outputs data either in metadata or AVI files. AVI formats use a "one frame in, one frame out" scheme. That means that the Presentation Time Stamp (PTS) and the Decoder Time Stamp (DTS) are the same, not allowing B-frames, as B-frames are constructed by using two frames at once, the previous and following I/P frame. Then, the encoding data will only present I and P frames. Table 4 shows the size of 16 frames taken from two consecutive GoP (the behavior is the same on other GoPs). As can be seen, Edge Detection I frames are reduced more than 50%. The reason is that after obtaining the edge, the Intra-frame coding performs better since the background is homogeneous. The results show that Intra-frame compression is a desired feature in these kind of devices.

The behavior of P-frames is totally different. In general, in MPEG encoding, a P-frame is encoded relative to the past reference frame. A reference frame is a P- or I-frame. The past reference frame is the closest preceding reference frame. Each macroblock in a P-frame can be encoded either as an I-macroblock or as a P-macroblock. An I-macroblock is encoded just like a macroblock in an I-frame. A P-macroblock is encoded as a 16x16 area of the past reference frame, plus an error term. To specify the 16x16 area of the reference frame, a motion vector is included. A motion vector (0,0) means that the 16x16 area is in the same position. This shows that Edge Detection decreases the amount of data to be sent. as the macroblock we are encoding. Other motion vectors are relative to that position. Motion vectors may include half-pixel values, in which case

Table 4. GoP sequence after encoding

Original		Edge Det.	
I	11507	I	5867
P	3996	P	5576
P	3272	P	103
P	3173	P	5801
P	2976	P	130
P	3117	P	5801
P	4755	P	111
P	2567	P	6028
P	3842	P	102
P	2924	P	5885
P	3770	P	113
P	3393	P	5754
I	11787	I	6112
P	3110	P	5594
P	3893	P	134
...	...	...	...

pixels are averaged. The error term is encoded using the DCT, quantization, and run-length encoding. A macroblock may also be skipped which is equivalent to a (0,0) vector and an all-zero error term. The search for good motion vector (the one that gives small error term and good compression) is the heart of any MPEG video encoder and it is the primary reason why encoders are slow.

We observe that the all of the original trace outputs' P-frames have the same order of magnitude. However using Edge Detection, we observe sequences consisting of a high-size P-frame (a few thousands of Bytes) followed by a short-size P-frame (few hundreds of Bytes). We think that the short-size P-frame may be due to small changes in the motion of the edge of the object. The high-size P-frame should encode again the whole edge. In any case, it should be taken into account that motion compensation is possible if simple algorithms can be used.

In AER [15], events are signaled when changes in pixel intensity reach a threshold voltage. AER uses 17 bits to encode each event. If N_{ev} is the number of event per frame, then $N_{ev}*17$ bits/event is the size of the frame. Edge Detection frames before compression have a size of 152064 bits/frame. AER obtain this sizes with 152064/17= 8944 events/frame. After compression, Edge Detection produces I and P frames of different sizes. The mean I and P frames in Edge Detection with 25 fps are 6050 and 3349 bits respectively. AER frames with 350 events/frame will produce frames of 5950 bits/frame. So AER with a number of events in the range of around 350 or 400 events produces the same amount of bits that Edge Detection with a complex coding as MPEG4. However, the advantage of AER imagers is that reduce the high compression overhead produced by encoders as is it is explained in [15].

4   Conclusions

Due to the bandwidth limitations and low power requirements of sensors, today every single aspect of a Wireless Multimedia Sensor Network is practically an open research area (like routing, MAC, network protocols). Unlike traditional sensor networks which transfer only small amount of data, visual data processing could be computationally very expensive simply due to the volume and information content of multimedia data. Hence, identification of traffic patterns could be one of the beneficial tools for working on such open areas. Thus, in order to design suitable networking protocols for wireless multimedia sensor networks, traffic should be characterized. These kinds of networks do not need to support streaming video instead of semantics of the sensed phenomenon which is the main function of the network (e.g., tracking objects in a certain physical environment). Since there are only limited-scope studies in the literature on this subject, it is extremely important to obtain a better understanding of the behavior of such traffic sources. In this study, "Object Recognition Techniques" (ORT) are considered as key components for reducing the amount of information to be sent to the sink since they could decrease the frame sizes. Moreover, coding techniques are further considered to reduce the temporal and spatial redundancies in frames.

Even though, there is no suitable encoder for this purpose today, using an unsuitable encoder such as MPEG4 gives an idea about effects of ORT. The results of our primary experiments suggest that using edge detection for scene decription can reduce the bandwidth utilization of the network significantly. However complexity of the algorithms should be handled carefully. Motion estimation makes the MPEG encoding algorithm slow due to the fact that it consumes more time and buffer size. It should be kept in mind that an encoder must satisfy the both mentioned requirements and operate in an optimal operating point by trading off the competing requirements. Nevertheless, data reduction with intelligent object and scene decription, without using complex algorithms, is the ultimate goal of our study.

References

1. Akyildiz, I.F., Melodia, T., Chowdhury, K.R.: A survey on Wireless Multimedia Sensor Networks. Computer Networks 51, 921–960 (2007)
2. Intanagonwiwat, C., Govindan, R., Estrin, D., Heidemann, J., Silva, F.: Directed Diffusion for Wireless Sensor Networking. IEEE/ACM Transactions on Networking 11(1) (2003)
3. Van Dam, T., Langendoen, K.: An adaptive energy-efficient MAC protocol for wireless sensor networks. In: 1st international conference on Embedded networked sensor systems (SenSys), LA, California, USA, pp. 171–180 (2003)
4. Rose, O.: Statistical properties of MPEG video traffic and their impact on traffic modeling in ATM systems. In: Proceedings of the 20th Annual Conference on Local Computer Networks (1995)
5. Fitzek, F.H.P., Reisslein, M.: MPEG-4 and H.263 video traces for network performance evaluation. IEEE Network 15(6), 40–54 (2001)

6. Chiasserini, C.F., Magli, E.: Energy Consumption and Image Quality in Wireless Video-Surveillance Networks. In: The 13th IEEE International Symposium on Personal, Indoor and Mobile Radio Communications (PIMRC 2002), Lisboa, Portugal (2002)
7. Feng, W.-C., Code, B., Shea, M., Feng, W.-C., Bavoil, L.: Panoptes: Scalable Low-Power Video Sensor Networking Technologies. ACM Transactions on Multimedia Computing, Communications, and Applications (TOMCCAP) 1(2), 151–167 (2005)
8. Boice, J., Lu, X., Margi, C., Stanek, G., Zhang, G., Manduchi, R., Obraczka, K.: Meerkats: A Power-Aware, Self-Managing Wireless Camera Network for Wide Area Monitoring. In: Workshop on Distributed Smart Cameras (DSC 2006), Boulder, CO (2006)
9. Rahimi, M., Baer, R., Iroezi, O., Garcia, J., Warrior, J., Estrin, D., Srivastava, M.B.: Cyclops: in situ image sensing and interpretation in wireless sensor networks. In: 3rd international conference on Embedded networked sensor systems (SenSys 2002), San Diego, California, USA (2005)
10. Teixeira, S., Culurciello, E., Park, J.H., Lymberopoulos, D.: Address-Event Imagers for Sensor Networks: Evaluation and Modeling. In: Fifth International Conference on Information Processing in Sensor Networks (IPSN), Nashville, Tennessee, USA (2006)
11. Zilan, R., Barcelo-Ordinas, J.M.: Object Recognition Basics and Visual Surveillance. Technical report: UPC-DAC-RR-XCSD-2008-1 (2008)
12. Video OCX, http://www.videoocx.de
13. Sikora, T.: MPEG Digital Video Coding Standards. IEEE Signal Processing Magazine (1997)
14. MPEG4-Parser, Video Trace Research Group, Arizona State University, http://trace.eas.asu.edu
15. Culurciello, E., Park, J.H., Savvides, A.: Address-Event Video Streaming over Wireless Sensor Networks. In: IEEE International Symposium on Circuits and Systems (ISCAS 2007), New Orleans, USA (2007)

Notes on Implementing a IEEE 802.11s Mesh Point

Rosario G. Garroppo, Stefano Giordano, Davide Iacono, and Luca Tavanti

Dip. Ingegneria dell'Informazione – Università di Pisa
Via Caruso, 16 – Pisa, I-56122 – Italy
name.surname@iet.unipi.it
http://netgroup.iet.unipi.it

Abstract. Wireless Mesh Networks (WMNs) are gaining wide popularity as a flexible and cost-effective access technology. Many vendors and network operators have already deployed their own proprietary solutions, and, in the meantime, the IEEE has set off Task Group 802.11s (TGs) to develop a common standard for WMNs. So far, however, TGs has not yet produced a final document and is still working on the draft. In this context, we have built a prototype mesh node as much compliant as possible with the (still unofficial) 802.11s draft. A software framework was developed using common off-the-shelf technology and deployed on top of the legacy 802.11 interface card. We believe that the availability of such a prototype will permit to readily test the features and the amendments to the draft as soon as they are proposed, thus returning immediate and significant feedbacks on their effectiveness. This is an important milestone, as simulation trials, though very useful, often do not give answers on the actual feasibility of the tested feature. And, in fact, the prototype already allowed us to experimentally evaluate the basic 802.11s characteristics, pointing out some shortcomings and suggesting possible improvements.

1 Introduction

Wireless Mesh Networks (WMNs) are rapidly gaining popularity as a flexible and cost-effective alternative to the traditional set of disjointed IEEE 802.11 Access Points (APs). In the typical 802.11 deployment, many APs are placed close to each other to extend the coverage range and capacity of the network. The operator needs to reach each AP with a cable, to coordinate the interactions between several adjacent APs and to manage user mobility. All these tasks obviously raise the operational cost of such a solution. WMNs are instead built around an all-wireless, self-organizing infrastructure, with the capability on carrying out those tasks in partial autonomy and the added advantage of being suitable for both residential premises and for hardly accessible places. As a consequence, many companies have already put on the market and deployed their own WMN solutions.

Though most of them are based on the common IEEE 802.11 MAC [1], these products are often not interoperable. Therefore, to harmonize these technologies,

L. Cerdà-Alabern (Ed.): Wireless and Mobility 2008, LNCS 5122, pp. 60–72, 2008.

in June 2004 the IEEE set up the 802.11s working group (TGs) whose goal was to define a standard architecture for WMNs. Unfortunately, the creation of the standard has turned out to be rather troublesome, as the first version of the draft received some 5,700 comments. Many issues are therefore still open, and proposals are still being submitted to TGs in order to complete and improve the upcoming standard (see [2]).

The effectiveness of the proposed solutions have been so far evaluated analytically or through simulations, but it is hard to find any experimental evidence of their goodness. This approach may lead to the definition of a standard that behaves differently from what meant and expected, because it may happen (not so infrequently) that the theoretical/simulation results do not closely match the practical working scenarios. As a corollary, we reckon that having a real testbed on which the the draft standard and its possible enhancements could be readily tested could greatly help the development of the standard.

All these reasons motivated us to start an implementation of a device that is as much compliant as possible with the current state of the IEEE 802.11s draft. In the paper, after a brief outlook of the draft, we describe the prototype we realised. However, rather than focusing on the single aspects of our prototype, we prefer giving more weight to the issues related to this implementation work. In our implementation, we mostly dealt with the path selection and frame forwarding facilities, which are two of the main innovations of 802.11s. Implementing them were very challenging tasks, as we had to face several issues that drove us to develop solutions not strictly adhering to the draft. Some issues were related to the available hardware and software tools (for example, modifying the basic 802.11 frames and creating the six-address Mesh Data frames was not feasible), while others came directly from the 802.11s draft in the form of missing, vague, or even faulty indications (mostly about frame forwarding and user station management).

2   Overview of IEEE 802.11s

In this section we provide just a short overview of the 802.11s main concepts. More details can be found in [3] or directly in the draft. Our implementation was based on version D1.02 of the unofficial draft [4], which was the one available at the time of our work, and therefore our overview is referred to that version. However, since the draft is rapidly evolving, some names and features have been slightly changed in the latest versions of the draft. Yet, the general architecture of the system and the major protocols and algorithms seems to be stable.

IEEE 802.11s builds on some already approved amendments to the standard, like 802.11a/b/g/n for the physical interface, 802.11e for accessing the medium and 802.11i for security, but it also conceives a new network architecture. In detail, the goal of TGs is "to provide a protocol for auto-configuring paths between stations over self-configuring multi-hop topologies in a Wireless Distribution System to support both broadcast/multicast and unicast traffic in a Mesh using the four-address frame format or an extension" [5]. Therefore, to accomplish its goal,

TGs introduces new mechanisms and frame formats for the configuration and operation of the Mesh network. By the way, it should be noted that 802.11s means to build the Mesh backbone on the already existing Wireless Distribution System, as defined in [1].

The basic 802.11s network entity is the Mesh Point (MP). Beyond having all the characteristics of a traditional 802.11 station (STA), every MP is also called to relay the traffic generated by other MPs to enable them to reach the intended destination through a multi-hop path. The set of connections among the MPs forms the wireless backbone of the Mesh. A MP can also have additional features, such as gateway/bridging functions, to connect to an external network, like the Internet. In this case, the MP is called Mesh Point Portal (MPP), or just Portal. A 802.11s network may have one or more Portals, and each MP chooses which Portal to use to get access to the external world. Another option for a MP is to give non-mesh devices, such as legacy client stations, access to the distribution system. In this case a MP becomes a Mesh Access Point (MAP) and must offer all the functions provided by the basic 802.11 standard. From the non-mesh STA point of view, the Mesh must be completely transparent. Figure 1 illustrates the relationship among the elements of a 802.11s mesh network.

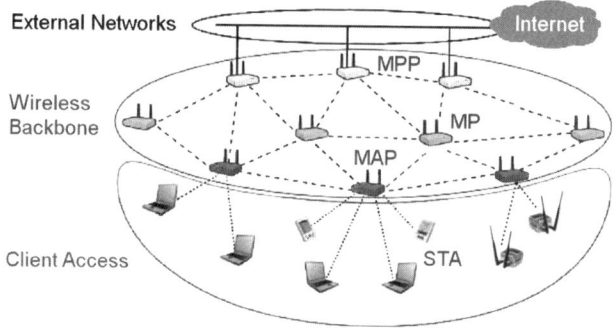

Fig. 1. Example of a IEEE 802.11s network

To build the set of paths that form the Mesh backbone, 802.11s brings the routing functions at level two of the OSI stack (with the name of "path se-lection"). This technique aims at making the Mesh functionally equivalent to a broadcast Ethernet, i.e. upper layer protocols see any intended destination within the Mesh to be directly connected at the link layer. The Hybrid Wireless Mesh Protocol (HWMP) is the mandatory path selection algorithm that all MPs must implement to guarantee the full functioning of the Mesh. HWMP combines an on-demand algorithm with a proactively built tree topology. The second tech-nique is used when at least one Portal is present in the network, since the tree shall be rooted at the Portal. Both the on-demand and the proactive techniques use common messages and processing rules. The Route Request (RREQ), Route Reply (RREP), Route Error (RERR) and Root Announcement (RANN) frames are flexibly structured to allow for the needs of both protocols. These messages are transported in the new Mesh Management frames.

The HWMP tree connects all MPs to the Portal (root), so that a path is always available towards the outside and between all MPs. It can be set up in two ways. The root broadcasts either Proactive RREQ (PRREQ) or RANN frames. The first technique aims at creating and maintaining a set of paths towards the root from all MPs. A MP receiving the PRREQ replies with a unicast Gratuitous RREP (GRREP), where it also inserts the address of its children MPs that use it to reach the root (dependent nodes). In this way the Portal is able to re-construct the full topology of the network. The second technique just disseminates information on how the root can be reached, leaving each MP the possibility to set up the path whenever it needs it (basically using the on-demand algorithm). Both PRREQ and RANN are re-broadcast by each MP.

The HWMP on-demand algorithm works like AODV [6]. A source Mesh Point S wanting to send data to a destination MP D broadcasts a RREQ frame indicating the MAC address of D. Two flags specify the handling policy of the frame: DO (Destination Only) and RF (Reply and Forward). If the DO flag is set, only the destination is allowed to reply to the RREQ, otherwise any intermediate node having a path to D can answer to S's request. If RF is set, intermediate nodes may reply to S, but shall nevertheless re-broadcast the RREQ. In any case, all nodes receiving the RREQ can add or update their path to S. Once D (or any allowed intermediate node) receives the RREQ, it sends S a unicast RREP. Intermediate nodes shall then forward this RREP to S along the best path, and, when the RREP reaches S, the path is set up and can be used for exchanging data.

Draft D1.02 adds new frame formats and modifies some of the existing ones. Most updates deal with the management of the Mesh services and algorithms. The new frames are the Mesh Management and Mesh Data, both of the Extended type. Both holds a Mesh Header, which consists of 4 or 16 bytes borrowed from the data field. In particular, the Mesh Header is used to store two extra addresses needed to forward the frames generated by user stations. The changes to the existing frames consists in adding new Information Elements (IEs) in the data field of the Management frames. Describing all the changes is out of the scope of the paper, so we point the interested reader to references [3] and [4].

3   Implemented Architecture and Features

We implemented a prototype 802.11s Mesh Point using common hardware (laptop and desktop PCs equipped with an IEEE 802.11b/g card) and open source software. We developed a software framework that was overlaid above the legacy 802.11 hardware and software (see Figure 2). Thanks to this solution we can employ the 802.11 Wireless Distribution System adding it the new Mesh services. Our framework can thus agree with the draft indications of having a transparent Mesh layer. In addition, the framework is structured in a series of modules, which allow to both testing different functionalities in several configuration options and easy upgrading to newer versions of the draft.

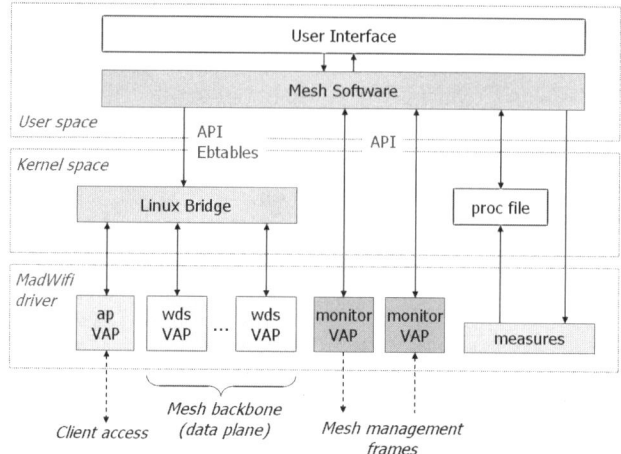

Fig. 2. Architecture of the implemented software framework

The PCs are run by the Linux operating system (Slackware 11, kernel 2.4). We then took advantage of the Multiband Atheros Driver for Wi-Fi (MadWifi) [7] to create several Virtual Access Points (VAPs). Each VAP is a separate instance of the same MadWifi driver. All VAPs work with the same physical device, but each VAP appears to the operating system as different wireless interface. Furthermore each VAP can be set up in a different operating mode. The mode determines the VAP behaviour and the set of processing rules the frames entering the driver undergo. In our implementation, several VAPs are set up in different modes in every MP. An *ap* VAP is used to interact with non-mesh devices (such as user stations) as defined in the basic 802.11 standard. This ensures that legacy stations are supported by the Mesh. The *wds* VAPs are necessary to form the WMN backbone. Each *wds* VAP builds a single point-to-point link to another MP. This is done using all the four addresses of the 802.11 frame header with the "from DS" and "to DS" flags set. Finally, two *monitor* VAPs are used to send and receive the new 802.11s frames. These frames can actually be built directly by the user (in the Mesh software layer) and are not processed by the driver when set up in the *monitor* mode. Therefore this mode can be exploited to handle the new Mesh Management frames introduced by the draft.

In every MP, the interfaces created by MadWifi are connected to each other by the Linux bridge. This module behaves exactly like a hardware bridge. In detail, every *ap* and *wds* VAP is connected to a port of the bridge, that forwards the frames between the interfaces according to their destination MAC address. Finally, Ebtables [8] has been used to filter and modify the frames traversing the MP. It is placed between the VAPs and the bridge, allowing processing the frames passing from one interface to the other (e.g. dropping, changing the source and/or destination addresses). In particular it has been used to virtually close the bridge ports when building the HWMP tree. The resulting framework is shown in Figure 2.

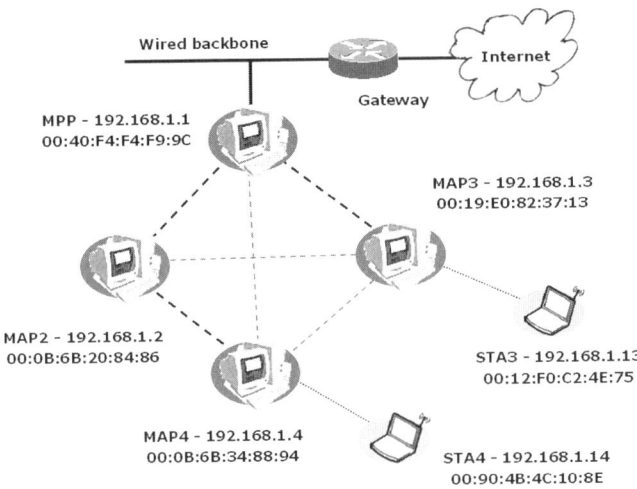

Fig. 3. The testbed network. Thicker lines represent the tree links.

The framework implements two functional levels: a management plane and a data plane. The former, run at user space, is made of all the procedures to build and maintain the Mesh. It exploits the *monitor* VAPs and controls the behaviour of the bridge (e.g. opening/closing the gates to build the backbone paths). The data plane works entirely at kernel space (Linux bridge and *ap* and *wds* VAPs) and handles the data frames; data frames are forwarded on the paths set up by the management plane.

3.1 Testbed Network

A simple test network was built with one PC set up as Portal, three PCs set up MAPs and two laptop PCs acting as legacy 802.11 user stations, i.e. non-mesh devices (see Figure 3). The Portal was equipped with proper routing capabilities, to separate the test network from the rest of the world at IP layer, and was then connected to the department Intranet, where a gateway provided Internet access.

When setting up the mesh, we configured our prototypes so that a HWMP tree is built through the PRREQ/GRREP messages and used as the main transport infrastructure, whereas on-demand paths are set up whenever a connection between stations is requested. Since the network topology is rather simple, we also built an extra custom module to give the mesh software pre-assigned link metrics. This was used in some scenarios to drive the MAPs to set up a particular topology.

4   Implementing Issues

During the development of our prototype, we had to face several issues. In this section we report the most interesting ones.

4.1   The New 802.11s Frames

The first point to discuss is that it was not always feasible to build and use the new 802.11s frames. We have seen in the previous section that Mesh Management frames were created by the Mesh Software module and used through the *monitor* VAPs. However, the new modified Data Frames could not be neither handled by the Mesh Software module (processing is too slow at user space), nor could be handled by the driver, because it would have requested a deep revision of the driver itself, which was out of the scope of our work. Therefore we had to find problem-specific solutions.

   Still about frames, though not properly an implementing issue, we noticed that the new Mesh services increase the use of broadcast frames. However, in 802.11 networks, many broadcasts may limit network performance, as these frames must be transmitted at one of the basic common rates [1], which is often lower than the rate used for unicast frames. So we devised a couple of solutions to reduce this drawback. Broadcast management frames produced by non-mesh stations are converted by the MAP into unicast frames addressed to the Portal. This is a sensible choice in most cases, as these frames are usually the tool to get some information from the Portal (e.g. ARP messages). Similarly, we constrained broadcast frames generated by the Portal to follow the HWMP tree. In this way all network functionalities have been retained, but the associated overhead was greatly reduced.

4.2   Station Management and the Proxy Mechanism

One of the main hurdles to overcome was the scarce weight that 802.11s gives to the fact that, in most cases, end-user stations are the main sources and sinks of data frames. For example, the format of the Mesh Management frame is designed for communications between MPs while the address extension fields (addresses 5 and 6) can only be used for forwarding frames involving non-mesh entities. As a result, managing non-mesh stations with frames not designed for this purpose was a very problematic situation.

   That is why we employed the frame format as defined in [9], which adds an extra address field to account for any station the MAP acts on behalf of. In particular, the "Proxied Address" field has been added to RREQ and RREP frames. When a RREQ is generated by a MAP because of a frame it received from an entity outside the mesh (e.g. a STA), this field is used to store the MAC address of that entity.

   Besides this feature, we found it convenient to integrate in our framework other Proxy mechanisms proposed in [9]. As a result, each MAP acts as a "Proxy" for its associated non-mesh station. It handles and converts the frames that enter or exit the Mesh, and it casts Proxy Update (PU) messages to inform the other MPs about its associated stations. Actually, the same 802.11s draft suggests that all MPs should store the address of every station associated to any MAP and the corresponding MAP address. However, this is just an informative note and no detailed procedures are given. On the contrary, we believe this feature is a key to perform many operations involving a non-mesh station (e.g. an on-demand path discovery).

A possible drawback of spreading such information across the mesh, is that it might become rather cumbersome, if each MAP broadcasts a Proxy Update (PU) frame every time a STA associates, disassociates, moves, etc. So we decided to put all the information in the Portal. The PU messages are sent by the MAPs directly to the MPP as unicast frames. This is coherent with the assumption that, since the final users of the Mesh (as an access technology) are principally consumers, most traffic is addressed to or comes from the Portal, which represents the gateway to the Internet. The root is then a privileged point for collecting and distributing information on the network.

4.3   The Airtime Metric

As for HWMP, to select the best paths the 802.11s draft defines the mandatory "airtime" metric (though other metrics can be used). The airtime metric is very similar to the well known ETX metric [10]. The definition is the following:

$$C_a = \left[O + \frac{B_t}{r}\right] \frac{1}{1 - e_{pt}} \tag{1}$$

In the formula, O is a constant that quantifies the channel access and protocol overhead (in terms of time), B_t is the test frame length (in bits), r is the transmission rate (in Mbps), and e_{pt} is the test frame loss ratio. MPs should continuously monitor and probe their links to keep the metric up to date with the current network state. These values are then exchanged with the neighbours through the LLSA (Local Link State Announcement) frames.

The airtime metric turned out to be very sensitive to changes in link usage. We noticed the occurrence of a perilous "ping-pong" effect among paths with similar metrics, especially under a "triangle" configuration. Figure 4 reports what happened in our testbed network. Accordingly to the HWMP tree, frames exiting STA follow the path C-B-R to reach the external network. However, it may occur that the unloaded (or just less loaded) path C-B-A-R momentarily offers a better metric, so MP B is induced to select that path to convey the traffic. Shortly, however, the new path (now more loaded) will suffer a metric degradation at the expenses of the old C-B-R path, which, on the contrary, is now unloaded. MP B is therefore bound to change again, thus creating an oscillating phenomenon. Since each path change brings with it some overhead, network resources may be significantly reduced.

We addressed the problem with a slight change in the metric (the transmission rate was replaced by the signal to noise ratio, SNR) to lessen the dependence on the frame loss percentage:

$$C_a = \left[k_1 + \frac{k_2}{SNR}\right] \frac{1}{k_3 - e_{pt}} \tag{2}$$

where k_1, k_2 and k_3 are constants used to balance the relationship between the SNR and the frame error ratio e_{pt}. For example, during our tests we found that

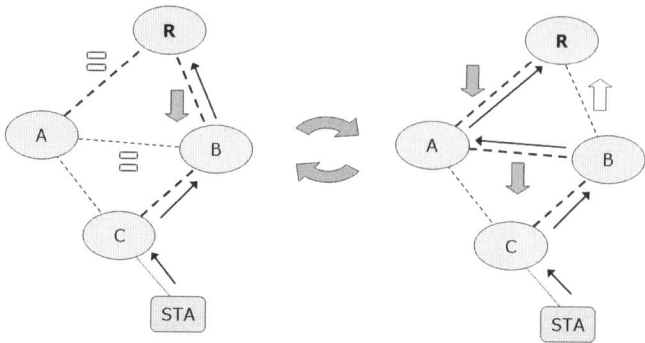

Fig. 4. The oscillating effect induced by the airtime metric

a good set of values is $k_1 = 3.2$, $k_2 = 250$, and $k_3 = 1$ (with SNR in dB). Then we also used a moving average, to smooth out the effects of sudden rises and falls in the metric, and forced the MPs to wait for a stable situation. A MP must receive a given number of messages announcing a better metric before switching to a new path. We set this number to be inversely proportional to the distance of the MP from the root. The reason is that the paths close to the Portal are usually the most loaded, hence changing them would have a greater impact on the network (service suspension and signalling overhead) than in a "leaf" link. We then verified that all these actions solved the instability and made the end-to-end paths more stable.

4.4   Optimizations to HWMP

We also optimized the on-demand procedure so that RREQ flooding is reduced. In particular, each node, instead of re-broadcasting all the received RREQs, merges them into a unique RREQ that is eventually sent out. This considerably lowers the protocol overhead, without reducing the protocol functionality. This choice is motivated by the fact that the first received RREQ is not always the one coming from the shortest/best path (access to medium is CSMA/CA, hence random delays occur). Therefore, after receiving a RREQ, a MP waits a short time to allow RREQs from different paths to be received as well. It then processes all the information and re-broadcasts only the RREQ announcing the best path. A similar criterion is used also when building the HWMP tree. A MP re-broadcasts only the PRREQ received from the neighbour closest to the root.

A further overhead reduction can be achieved for the GRREP messages. 802.11s draft dictates that the complete list of all dependent nodes is inserted in the GRREP. However this information is redundant, as many copies of the same list reach the Portal. Hence we put in the GRREP just the direct children of the MP. This also allowed a more precise knowledge of the topology of the Mesh.

5   Validation Test

The functionality of our prototype was tested in a series of network scenarios. Please be aware that, being the development still at the prototype phase, we mostly aimed at verifying the correct behaviour of the system, rather than extracting performance figures (which nonetheless might be the goal of a successive measurement campaign). In detail, we tried its capability of building the HWMP tree, creating on-demand paths and connecting a client to the external network or to another client. In this section we report just an example, based on the network in Figure 3.

In the test we asked the mesh to find a path for connecting two clients, STA3 and STA4. After the initialization phase, during which the tree was formed (the two branches are MPP-MAP2-MAP4 and MPP-MAP3), we launched a ftp transfer on STA4 towards STA3. The application required at first to get the MAC address of STA3, whose IP is known. In practice, STA4 started an ARP procedure. The subsequent frame exchange is reported in Figure 5, where the address fields of all frames are also shown. At first, STA4 broadcast an ARP request message, which was received by MAP4. This node transformed it into a unicast frame with final destination the Portal (in accordance to what exposed in 4.1). The frame was thus sent to MAP2 that forwarded it to MPP. The Portal, being aware of all stations in the network, replied with STA3's MAC address. The ARP reply message, addressed to STA4, reached MAP2, then MAP4 and finally STA4.

Once STA4 got knowledge of the association IP-MAC of STA3, could start transmitting data frames addressed to MAP4 but with final destination STA3 (in accordance to the standard). When MAP4 received such frames, having no direct "route" to STA3, it could only forward them to MPP. The data frames therefore moved along the tree towards the MPP and then back on the other

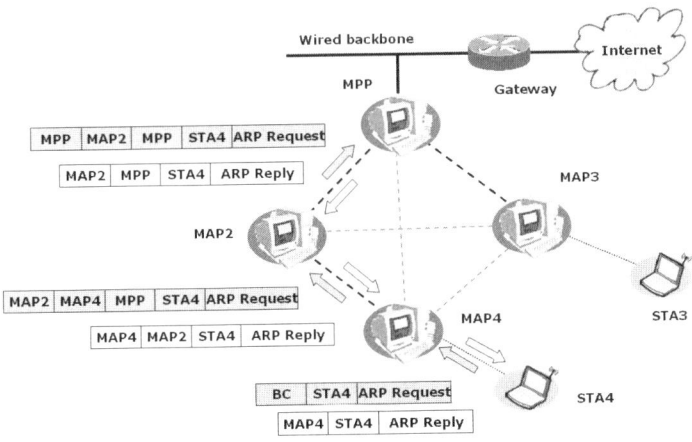

Fig. 5. The frame exchange sequence related to the ARP procedure. Address fields are in this order: receiver (RA), transmitter (TA), destination (DA), source (SA).

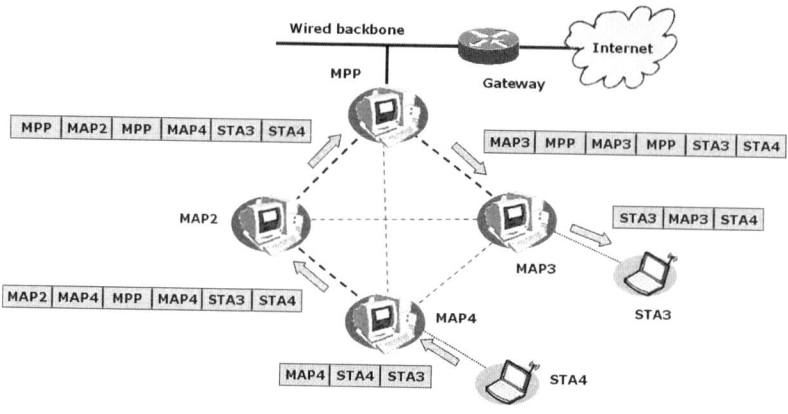

Fig. 6. The address fields of a data frame from source (STA4) to destination (STA3)

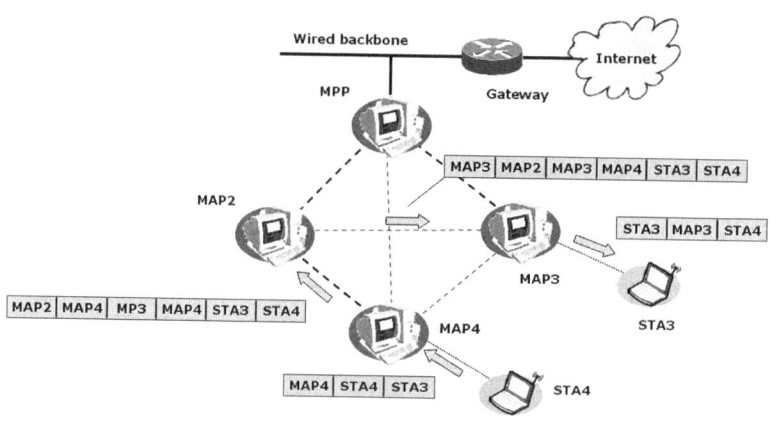

Fig. 7. The final path from STA4 to STA3

branch to MAP3 and then STA3. In figure 6 we report again the situation of the data frames as they traverse the mesh. In this case the use of the six-address format is evident. Fields five and six holds the non-mesh stations which are the source and destination of the end-to-end path.

While sending data frames to the Portal, however, MAP4 also started a path discovery procedure, broadcasting a RREQ message to set up a direct path to STA3. The RREQ message reached MAP3, which performed its duty as a proxy and replied with a RREP on behalf of STA3. When MAP4 received the RREP it immediately started sending the new frames on the newly built path (in our case MAP4-MAP2-MAP3). Figure 7 shows the new path and the address fields of the data frames. Note that this is not the shortest path, but the most convenient according to the metric values we imposed on the links with our custom extra module (see 3.1).

6   Conclusions

In the paper we presented some notes we could draw from our implementation of a prototype IEEE 802.11s Mesh Point. The prototype was as much compliant as possible with the current draft. The implementation was realised with common tools, such as Linux-based PCs and open source software. A mesh software framework was implemented on top of the network interface card driver, in accordance to the standard dictate of being the Mesh just a logical interface. The framework is also flexible enough to allow an easy upgrade to the newer editions of the draft (as soon as they become available). The functioning of our solution was successfully verified in a testbed network.

Some useful considerations can be drawn from this work. At first, the availability of a prototype 802.11s network allowed us to readily test the features and the amendments to the draft almost as soon as they are proposed. Immediate and significant feedbacks on their effectiveness were returned, as our prototype was able to spot some flaws.

The suggested airtime metric, used to build the tree paths to the Portal, may be the cause of instability problems. A fine tuning of the path selection parameters is therefore advisable, but seeking other metrics is probably an even better option. Also, HWMP itself can be ameliorated. For example we proved that PRREQs flooding can be reduced and GREEPs can be made lighter without detriment for the full functionality of the Mesh.

Finally, the management of non-mesh stations needs to be improved. This can be done by enhancing the use of Proxy mechanisms at the MAPs, and by introducing more complete frames. Moreover, when a HWMP tree is present, letting the Portal have knowledge of all STAs is another feasible and convenient feature. As a side remark, the goodness of these findings was confirmed by the latest changes to the draft, as they have been included in one of the latest versions [11].

Acknowledgements

This work was supported by the Italian MIUR-FIRB project "Integrated System for Emergency" (InSyEme), under grant number RBIP063BPH. The authors wish to thank Giorgio Barsacchi for his useful contribution in the implementation of the described prototype.

References

1. IEEE Std 802.11-2007 (Revision of IEEE Std 802.11-1999), IEEE Standard for Information technology – Telecommunications and information exchange between systems – Local and metropolitan area networks – Specific requirements – Part 11: Wireless LAN Medium Access Control (MAC) and Physical Layer (PHY) Specifications (June 2007)
2. IEEE P802.11 Task Group S – Status of Project IEEE 802.11s – Mesh Networking, http://grouper.ieee.org/groups/802/11/Reports/tgs_update.htm

3. Hiertz, G.R., Max, S., Zhao, R., Denteneer, D., Berlemann, L.: Principles of IEEE 802.11s. In: Proc. of 16th International Conference on Computer Communications and Networks (ICCCN) (August 2007)
4. IEEE P802.11 Task Group S, IEEE Unapproved draft P802.11s/D1.02 (March 2007)
5. IEEE 802.11s Project Authorization Request (PAR) (August 2007)
6. Perkins, C., Belding-Royer, E., Das, S.: Ad hoc On-Demand Distance Vector (AODV) Routing, IETF RFC 3561 (July 2003)
7. Multiband Atheros Driver for Wi-Fi, http://madwifi.org
8. Ebtables, http://ebtables.sourceforge.net
9. Gossain, H., et al.: Proxy Frame Forwarding. IEEE document 802.11-07/0337r1 (March 2007)
10. De Couto, D.S.J., Aguayo, D., Bicket, J., Morris, R.: A High-Throughput Path Metric for Multi-Hop Wireless Routing. In: ACM Mobicom (2003)
11. IEEE P802.11 Task Group S, IEEE Unapproved draft P802.11s/D1.08 (January 2008)

Interference-Aware Channel Assignment
in Wireless Mesh Networks

Rosario G. Garroppo[1], Stefano Giordano[1], Davide Iacono[1], Stefano Lucetti[2],
and Luca Tavanti[1]

[1] Dept. of Information Engineering
University of Pisa, Italy
{name.surname}@iet.unipi.it
[2] NetResults S.r.l., Pisa, Italy
stefano.lucetti@netresults.it

Abstract. The emerging IEEE 802.11-based Wireless Mesh Networks (WMNs) suffer of poor performance in terms of throughput. The use of multi-hop paths to relay the traffic raises the problems of intra-flow and inter-flow interference in the wireless backbone. A possible solution to this problem is exploiting a multi-channel framework. In this context, we develop an innovative interference model that permits to take into account the actual signal power received from each interference source, the interference produced by sources external to the WMN, the link utilization factor and directive antennas. Starting from this new model, we propose a channel assignment algorithm for multi-radio WMNs which is based on graph theory. Its performance is evaluated through simulations, which show a great improvement over existing channel assignment algorithms.

Keywords: wireless mesh networks, multi-channel, interference modelling, graph theory, radio resource management.

1 Introduction

The increased popularity and the growth in the number of deployed IEEE 802.11 Access Points (APs) have raised the opportunity to merge together various disjointed wireless LANs to form a unique Wireless Mesh Network (WMN). In this architecture only few nodes, called Portals (MPPs), interface to the external world thanks to their bridging/gateway functions, while several Mesh Points (MPs) form a wireless backbone acting as relay nodes. At the same time, some (or all) the MPs are able to provide network access to user devices (stations) that are inside their coverage range. Such MPs are then called Mesh Access Points (MAPs)[1]. As a result, WMNs are supposed to handle a large amount of traffic, including flows that demand specific requirements, such as VoIP calls or multimedia broadcasting.

---

[1] These names are essentially those defined by the upcoming IEEE 802.11s draft standard. However the network architecture is general enough to fit a wide range of WMNs.

L. Cerdà-Alabern (Ed.): Wireless and Mobility 2008, LNCS 5122, pp. 73–88, 2008.
© Springer-Verlag Berlin Heidelberg 2008

To this end, one serious limit of 802.11-based WMNs is the fact that the IEEE standard plans to use the CSMA/CA access protocol over a single frequency/channel [1]. In the Mesh architecture, the MPs shall transmit their traffic and, simultaneously, forward what they receive from other MPs, thus forming a series of multi-hop paths that constitutes the wireless backbone. The weave and intense use of such paths give rise to inter-flow and intra-flow interference. This phenomena can easily grow to unbearable levels, causing the network to offer poor performance in terms of throughput, delay, and jitter [2].

Let us consider, for example, the network topology in Fig. 1, left. The packets of a client (STA1) that wants to establish a connection to the Internet follow a certain path through the network (Flow1). Along this path, each MP has to forward the packets of STA1, but, due to the use of CSMA, it often senses the carrier busy because of the transmissions of other MPs on the same path. It has then to wait the end of the other transmissions before accessing the medium and transmitting its packets. Transmissions of the same flow interfere with each other.

Moreover, if there is another client (STA2) that communicates with a device in the Internet, the MPs on the second path (Flow2) will sense the carrier busy due to the transmissions of the MPs on Flow1 and vice versa (see Fig. 1, right). The interference among the two flows thus boosts the problems of reduced throughput and increased latency for the single flow.

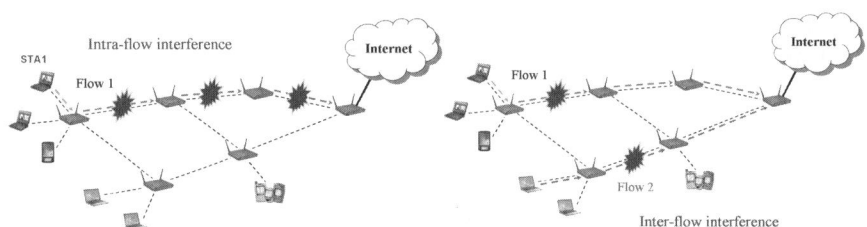

Fig. 1. Example of intra-flow (left) and inter-flow (right) interference

In this scenario, the use of multi-channel is more than just an option, rather it becomes a necessity if network capacity is a key design issue. The utilization of multiple channels that are assigned to the various wireless backbone links allows multiple transmissions to exist at the same time, reducing medium access times and increasing end to end throughput.

Many channel assignment algorithms and protocols have already been proposed to solve this problem at physical and/or data link layers. A distributed channel assignment solution has been presented in [3], while a distributed assignment joined with routing has been developed in [4]. A centralized approach was instead the focus of [2] and [5], which exploited graph theory. References [6] and [7] merged routing with centralized channel assignment. The above solutions do not foresee a 802.11 MAC layer modification, whereas other works that utilize channel switching on a single radio do need it [8][9].

By the way, IEEE 802.11b/g and 802.11a provides 3 and 12 (under ETSI regulatory domain for outdoor use) non overlapping (frequency) channels, respectively. However some works prove that there is still a certain amount of interference due to signal pulse tails overlapping. In [10], authors show that in 802.11b/g only channel 1, 6 and 11 are quite separated in frequency. On the other hand, [11] reports that two 802.11a channels are quite isolated among themselves if their spectral distance is greater than one channel bandwidth. In this configuration they interfere with each other very weakly and network performance is not compromised.

We propose a centralized channel assignment algorithm for multi-radio WMNs that accounts for a much more detailed modelling of the interference. Differently from previous works, we consider a physical interference model, developed according to radio propagation rules, to account for the real signal power received from each interference source. In fact, the common protocol model used in most previous works assumes binary interference: a node is either within or out of the interference range of another. There is no difference between a node that is very close to another and one on the edge of the interference range, which could actually be very large. They are considered to interfere with it in the same manner. The path loss experimented by the signal is generally not taken into account. Clearly these assumptions drive to a rough estimate of the interference, while a more realistic computation can provide for a much finer tuning of channel assignment.

In addition to the estimation of the interference among MPs, the strength of our channel assignment scheme is that many other interference factors are accounted for. Shortly, interference coming from external sources, like other WLANs or devices working in the same frequency band (e.g. radars), but not being part of the WMN, is considered. Furthermore, interference that can arise among radios on the same node is taken into account as well. Indeed, due to path loss absence between source of interference and the affected transceiver, this factor may be even more critical than the others, up to complete disruption of service. Finally, other degrees of freedom related to the WMN design and implementation are included, such as the possibility that a node is equipped with directional antennas (as in point-to-point links or point-to-multipoint sectors) besides the classical omni-directional ones.

The rest of the paper is organized as follows: section 2 provides a classification of channel assignment algorithms; in section 3 we present the network model we used in this work; in section 4 we describe the proposed channel assignment algorithm and in section 5 we report the simulation results. Finally, conclusions are reported in section 6.

2   Classification of Channel Assignment Schemes

Channel assignment consists in setting the MP radio interfaces to certain channels in order to minimize interference. Channel allocation algorithms can be classified considering the time scale on which the assignment occurs [6][12].

Static Assignment
In this approach a channel is assigned to a certain interface permanently or for a very long time. Static assignment techniques can be further divided into two sub-categories.

- **Common Channel Assignment (CCA):** In CCA subcategory [13] the radio interfaces of every node are assigned to the same channel subset. The main advantage of this kind of approach resides in the fact that the connectivity degree is the same as the single-channel approach. However, as a consequence, there is an increase of co-channel interference comparing with other methodologies.
- **Varying Channel Approach:** In this approach the radio interfaces of different MPs are set to partially overlapped channel subsets. The frequency diversity assures a reduction of the interference experimented by the MPs, but not the connectivity preserving as in CCA leading to possible network partitions. With this approach, there is a possibility that the length of the routes between nodes may increase. To this purpose it is important to perform this allocation very carefully in order not to change the network topology. The algorithms proposed in [5] and [2] utilize this approach.

Dynamic Assignment

In this approach a radio interface can frequently switch from one channel to another, even on a per-packet basis. When this approach is used, two nodes need to be carefully synchronized to ensure that both of them are communicating on the same channel at the same time. To this end the nodes may visit a common "rendezvous" channel periodically. In [4] a distributed algorithm utilizes only local traffic load information to dynamically assign channels and to route packets. In the Slotted Seeded Channel Hopping (SSCH) [8] each node switches channels synchronously according to a pseudo-random sequence, in order to let neighbours meet on the same channel, similarly to frequency hopping schemes. In Multi-channel MAC (MMAC) [9], every node needs to be synchronized, so that each beacon starts at the same time. At the start of each beacon interval, every node listens on a common channel to negotiate channels in a certain window time. Then, nodes switch to their agreed channel and exchange messages on it for the rest of the beacon interval. This kind of approach utilizes all the available channels, but on the other hand it imposes strict constraints on synchronization and switching delay, which may require specific hardware and firmware.

Hybrid Assignment

This approach combines the two exposed before, assigning to some radio interfaces a channel for a long time while dynamically commuting the others. This strategy has the advantage to require simplified coordination algorithms while retaining the flexibility of dynamic assignment. Hybrid strategies can be further classified depending whether the interfaces that apply static assignment use a common channel approach, or a varying channel approach. The scenarios presented in [3] and [6] utilize this approach. Again, as for dynamic assignment, this approach has the constraints on synchronization and switching delay due to dynamic allocation of non fixed interfaces.

3   Problem Formulation

Our algorithm is based on the assumption of a tree-topology wireless backbone (which is the one usually created in IEEE 802.11s WMNs [14]), where each node can be equipped with several network interfaces. The intended solution exploits a centralized approach like in [5], where a central server performs the actual channel assignment. It works independently from the specific radio technology of a mesh node. In this manner we avoid any MAC modification that would be very troublesome and may lead to interoperability problems. Moreover, we do not address joint channel assignments and routing issues, because route changes may occur on a much shorter time scale than frequency assignments, thus leading to excessively complex assignment problems. We formulate the channel assignment as a mesh topology control problem. In our allocation we also want to preserve the connectivity among the MPs.

In order to properly account for the interference in the channel assignment problem, we base our solution on the use of the graph theory [15]. In particular, a connectivity graph is used to model the network topology, while a weighted conflict graph [16] is built to consider the various links in the wireless backbone that interfere with each other.

We model the WMN topology by means of a bidirectional graph, called *connectivity graph*, where each MP is represented by a vertex and each wireless link is represented by an edge. Two vertices i and j are connected if and only if $d_{ij} \leq \min(R_i, R_j)$, where d_{ij} is the distance between i and j, and R_i and R_j are their transmission ranges. This definition is obviously rather conservative, as it results in a rather "simplified" graph. Other definitions could be used, for example setting R_i as the sensing range and $d_{ij} \leq \max(R_i, R_j)$. Such definition will then yield a more connected graph, but, as it will be explained later, a much more complex conflict graph. Hence, for the sake of simplicity and without loss of generality, we will keep the first definition.

We then model interference by means of a *conflict graph*. In this graph each vertex represents a wireless link in the connectivity graph and two vertices are linked to each other by an edge if their corresponding links are interfering. Fig. 2 and Fig. 3 report an example WMN topology represented by a connectivity graph and the related conflict graph. The increase in the complexity of the conflict graph is apparent.

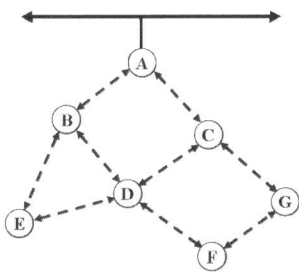

Fig. 2. Example of Connectivity Graph

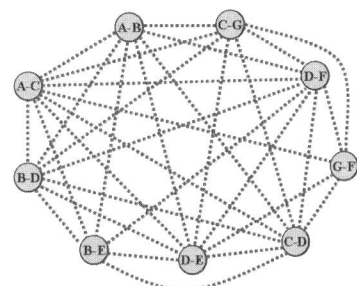

Fig. 3. Example of Conflict Graph

In order to accurately consider the interference among the wireless links, a weight is assigned to each edge of the conflict graph. These weights depend on the underlying interference model. In literature there are two main models: the protocol and the physical models [16].

Protocol Model: Let i, j, $1 \leq i,j \leq N$, denote the nodes of a plane wireless network and d_{ij} the distance between them. Each node has a radio interface that has a communication range R_i and a potential interfering range R_i'. In the protocol model a communication is successful if the following conditions are both verified:

1. $d_{ij} \leq R_i$ and $d_{ij} \leq R_j$
2.   Any node k, such that $d_{kj} \leq R_j'$ and $d_{ki} \leq R_i'$ is not transmitting

If any of these conditions is not true, the communication fails. As a result, the weights in the conflict graph are binary, because a node either produces a destructive interference or no interference at all. For example, the link (u,v) in Fig. 4 is interfering with link (i,j) because node v is within the interference range of node i.

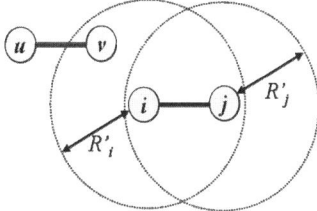

Fig. 4. Interfering links according to the Protocol Model

Physical Model: Using the same notation as before, in this model a communication is successful if the $SINR_{ij} \geq SINR_{th}$, where $SINR_{ij}$ denotes the signal-to-noise plus interference ratio at node j for the transmission received from node i. In the $SINR_{ij}$ all signal contributions from all nodes are considered, each one with its received strength (path loss is accounted for). Therefore the weights in the edges of the conflict graph vary according to the physical distance and the activity of the other nodes.

Assuming binary interference means that a node very close to another interferes with it in the same manner as a node on the edge of the interference range, which could actually be very large. Though these assumptions may make the problem solving easier, they can also lead to an imprecise solution of the channel allocation challenge.

3.1 Interference Model

In order to accurately estimate the interference, we start from the physical model and we extend it with additional features.

As seen above, radio propagation rules [17] are used to compute the real signal power received from each interference source. However, the physical model only accounts for interference coming from others MPs, whereas in a multi radio scenario the interference from the other radio modules installed on the node itself may also contribute significantly. In fact, even though the best isolation techniques are in place these sources are not subject to path loss and may affect the receiver with a signal power comparable (or even stronger) to the one coming from the other nodes. To account for this factor we cannot make use of the radio propagation rules, which are valid in the far field only. Rather, we must refer to the coupling phenomena and isolation figures. Spectrum overlapping issues must also be considered.

Moreover, not only the inter-MP and intra-MP interference is considered, but also the interference that comes from sources external to the mesh, like other WLANs or devices working in the same frequency band (e.g. radars). These sources may make some channels unusable in some parts of the mesh network, thus putting an extra constraint to the channel allocation problem.

We also include the possibility that a node is equipped with directional antennas (besides the classical omni-directional ones). The directionality of the antenna has an impact on both the logical topology of the network (i.e. the connectivity graph) and on the weights of the edges in the conflict graph.

Finally, we account for the traffic pattern, which determines the duty cycle of each radio interface of each node, and consequently gives a measure of how much a node is interfering with the others because it is transmitting. Utilization factors are used to consider the amount of time a node is in the transmitting phase.

These concepts can be formalised as follows:

- Let $\{i, | 1 \leq i \leq N\}$ be the set of MPs (nodes) in the network and (i,j) denote a link between the node i and node j;
- Let $G(V,E)$ be the connectivity graph, where $v \in V$ are the nodes (vertices), and $e \in E$ the wireless links (edges);
- Let $IE(i,j)$ be the set of neighbours of node i minus j and the neighbours of node j minus i;
- Let $H(L,S)$ be the conflict graph, where $l \in L$ are the wireless links (vertices) and $s \in S$ the edges that indicate interference between them;
- Let $C = \{c \mid c = 1,2,\ldots,K\}$ be the complete set of channels and C_n the subset of channels available at node n;
- Let $\beta(c_l,c_g)$ be a factor that accounts for the spectrum overlapping between channels c_l and c_g;
- Let $m \in \{1 \ldots M_n\}$ denote an interface of node n and $S_m^n(p)$ be the power received at the mth interface of node n from the interference source p (including the antenna gain in the direction of p);
- Let ρ_m^n be the utilization factor of the interface m on node n, which quantifies the fraction of time the interface is in the transmission state. In ideal and non saturation conditions, the sum of all the ρ in a neighbourhood should be no greater than one.

The procedure to build the conflict graph is described below. Let us take into account a generic wireless link (i,j) between nodes i and j (Fig. 5). We consider as interfering sources of node i all its neighbouring nodes (i.e. the nodes to which an edge exists in the connectivity graph) minus j (node u and v in the example) and all the nodes towards which the i's neighbours have a wireless link (i.e. an edge). The same is done for node j. The resulting set $IE(i,j)$ therefore accounts for all potential

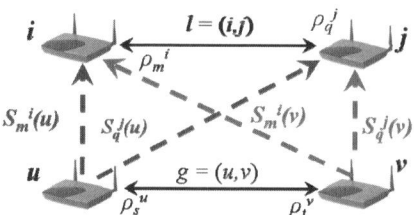

Fig. 5. Example of Interfering nodes

interferers of link (i,j). Since these interfering nodes have wireless links to other nodes, we consider all these links as interfering with (i,j), and, as a result, they are connected to the $(i,j) = l$ in the conflict graph.

Considering the example of Fig. 5, the weight assigned to the edge that connects g and l in the conflict graph is evaluated as follows. Assume that link $l = (i,j)$ is operating on channel c_l. For link $g = (u,v)$, operating on channel c_g, that interferes with l, we evaluate the power received by i and j from u and v. This can be done according to any radio propagation model is deemed appropriate for the specific scenario (e.g. indoor, outdoor). For example, we can use the shadowing model [17]:

$$S_m^i(u) = \frac{P_t G_t G_r \lambda^2 d_0^{n-2}}{(4\pi)^2 d_{iu}^n} \tag{1}$$

The formula refers to the mean power received by the mth radio module of i from u. Similar formulas apply in the other three cases (i from v, j from u, j from v). In the formula, P_t and G_t are the transmission power and the antenna gain of node u, G_r is the antenna gain of node i, d_{iu} is the distance between the two nodes, d_0 is the reference distance (normally set to 1 meter for indoor scenarios), λ is the wavelength and n is the path exponential loss. The four $S(\cdot)$ contributions are then weighted with the spectrum overlapping factor $\beta(c_l, c_g)$ and with the utilization factors of both the interfering and the source nodes. Thus, for the example in Fig. 5, we obtain the following expression:

$$W_{l,g}(c_l) = \beta(c_l, c_g)[(1 - \rho_m^i)(\rho_s^u S_m^i(u) + \rho_t^v S_m^i(v)) + (1 - \rho_q^j)(\rho_s^u S_q^j(u) + \rho_t^v S_q^j(v))] \tag{2}$$

This formula yields the weight $W_{lg}(c_l)$ that is associated to the edge (l,g) in the conflict graph. It is important to observe that the terms $1 - \rho_m^i$ and $1 - \rho_q^j$ are used to consider the fraction of time the nodes i and j (more exactly, the mth interface of i and the qth interface of j) are in the receiving state, as this is the time they are subject to destructive interference. Similarly, we also take into account the utilization factor of u and v (ρ_s^u and ρ_t^v respectively), which are the fraction of time they may cause interference. The utilization factors allow us to consider the traffic that flows into the network, so that we have a very accurate knowledge of the average channel conditions. We assume that the traffic pattern is known (e.g. the central coordinator may also have knowledge of the routes and the average amount of data per flow). By means of $\beta(c_l,c_g)$ we may also appraise the interference received in the frequency band of channel c_l from a signal on channel c_g [10][11].

It is worth noting that the term $W_{l,g}(c_l)$ depends on many factors: besides the received interfering power from external sources, as in the classical physical model, we also account for the overlapping factor between the channels and on the traffic that flows into the network. In this way, we have a deep knowledge of the channel conditions and, therefore, we can have a better estimation of the interference the MP experiences.

As already mentioned, a further, and non-negligible interference factor, is the power received from the radio interfaces that are on the same board of the node under consideration. This concept can be formalised with the factor $\gamma^l_{m,p}(c_m,c_p)$, which measures the interference on interface m (which works on channel c_m) of node n due to couplings with interface p working on channel c_p. The actual value of this term can be retrieved from either the device datasheet or from experimental measurements. The sum of the interference from all on-board interfaces can be expressed as:

$$\Gamma_n(c,m) = \sum_{p \neq m} \gamma^n_{m,p}(c_m,c_p) \tag{3}$$

4   Channel Assignment Algorithm

The problem of optimal channel assignment in an arbitrary mesh topology has been proven to be NP-hard based on its mapping to a *graph-colouring* problem [18]. For this reason our approach is heuristic. The proposed algorithm integrates the one proposed in [5] with the Breadth First Search (BFS) exploration priority used in [2]. We assume that the largest amount of traffic comes from or is directed to an external network, so the gateway and the nodes close to it are the most loaded. From this point of view, it seems sensible to give these nodes a higher priority in the channel allocation scheme. We achieve this goal by means of BFS.

We also assume that network conditions are quite stable and do not change significantly over time (at least during the allocation computation and set up). However, possible changes are accounted for by means of periodical refresh of the channel allocation solution, which may occur due to an alteration of the traffic pattern (e.g. following to varying routes) or the propagation conditions (e.g. new fixed obstacles).

The key of the assignment problem is to allocate channels to interfaces in order to disconnect the conflict graph vertices as much as possible; ideally the graph will be completely disconnected after the channel allocation.

For each link $l = (i,j)$ between nodes i and j, we compute the function f_l with the objective of finding the channel $c \in C_{ij} = C_i \cap C_j$ that minimizes the interference on link l due to all interfering links and to the internal sources of both nodes i and j:

$$f_l = \min_{C_{ij}} \sum_g W_{l,g}(c) + \Gamma_i(c,m) + \Gamma_j(c,q) \tag{4}$$

In the formula, g represents a generic interfering link that is connected to l in the conflict graph and $W_{l,g}(c)$ is the weight assigned to the edge between l and g, as

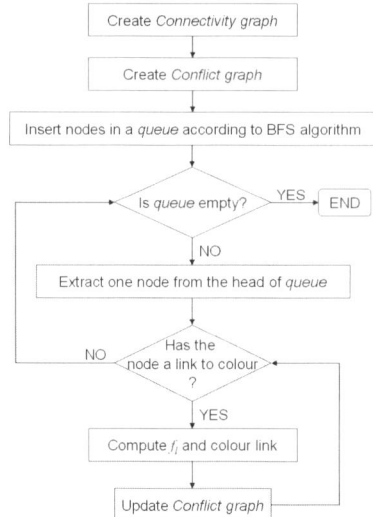

Fig. 6. Flowchart for the proposed algorithm

evaluated according to (2). The function f_l is then computed for all links l in the connectivity graph, until all links are assigned a channel. The flowchart for the complete channel allocation algorithm is shown in Fig. 6.

After creating the connectivity graph and the conflict graph (as described before), the nodes are visited according to BFS [15]. BFS works on the connectivity graph and, as already mentioned, we assume the network has a tree topology, with the Portal being the root. Starting from the root, we insert all nodes connected to it (level one children) in a list of nodes to be visited (denoted as *queue*). These nodes may be ordered (the criterion is irrelevant). Then, for each of these nodes, all neighbours (level two children) are added at the bottom of the list (as long as they have not been added before) and ordered. The same procedure is repeated until all the nodes in the tree have been added.

Once the queue of nodes is filled, the nodes are visited one by one and, when a node has some link to be coloured (i.e. a channel must be assigned to that link), the algorithm computes function f_l and finds the channel c that minimizes the interference for the considered link (at the beginning we assume that all interfaces are momentarily tuned on the same channel – the most unfavourable case). The resulting channel is definitively assigned to that link. The conflict graph is then updated, and, following to the allocation, some vertices are presumably disconnected. In case a node has some links yet to be coloured but all its radios have been definitively tuned (i.e. the number of assigned channels equals the number of radio interfaces), the algorithm will reuse one of the already set channels. In this context, it is important to account for directional antennas, as some channels may be in use on interfaces that do not guarantee the destination to be in the coverage field. The algorithm then proceeds to compute f_l for the next link and then for the other nodes, until all links have been assigned their final channel (or, in other words, all edges of the connectivity graph have been coloured).

5   Simulation Results

The efficiency of the proposed channel allocation algorithm merged with the described interference model is assessed in terms of traffic throughput and end-to-end delay. We have compared the performance of the proposed channel allocation with other solutions such as CCA and CLICA [5]. In CCA all the nodes have the same channel subset. Similarly to our algorithm, CLICA uses a conflict graph, but with binary weights, according to the protocol interference model (see Section 2).

We simulated the scenario shown in Fig. 7. In our simulations no external interference sources are present, so that each node can utilize all the available channels. Furthermore no directional antennas have been used. Though these two features are part of our algorithm, we decided not to take advantage of them in the simulations to allow for a fair comparison with the other schemes, which do not provide for them.

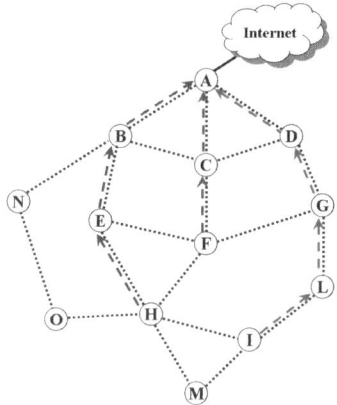

Fig. 7. Simulated Network

According to our assumption that the majority of the traffic is directed to or comes from an external network, we have activated three flows, all of them directed to the gateway (node *A*). In this manner the MPP will be heavy loaded. In particular the traffic sources are nodes *H*, *F* and *I*. The dashed arrows show the routes that the packets follow. The traffic sources are CBR (Constant Bit Rate) and the transport protocol is UDP (User Datagram Protocol). The choice of such a protocol is justified from the fact that a TCP (Transport Control Protocol) connection has inside a congestion control that hides the potentiality of the multi-channel communication, thus flattening the performance of the different channel allocation algorithms.

In order to simulate such a scenario, we took advantage of the ns-2 simulator. More precisely we used the ns-2.30 release and applied the patch developed in [19] to manage the multi-radio architecture. It supports several interfaces on a node each attached to a channel/link. On the other hand, the utilization of multiple channels refers to a single routing/forwarding or MAC entity deciding which channel to use. From this point of view channel overlapping is not considered. In our simulations we used the IEEE 802.11b MAC layer, 11 Mbps for data frame and 1 Mbps for control frames, with omni-directional antennas and 12 non overlapping channels. The reasons of this choice are mainly two: firstly, we wanted to compare the results that come from previous works which utilize exactly this configuration [5]. Secondly, we suppose that channel allocation makes sense when a large amount of channels is available. For this reason we employed 12 non overlapping channels (as it might be a working scenario based on IEEE 802.11a). Moreover ns-2 does not consider physical channel overlapping, but just logical separation. However, following the results provided by [11], we avoided that interfaces on the same node are assigned to adjacent channels. This was achieved by setting the $\gamma''_{m,p}(c_m,c_p)$ terms of equation (3) to one for adjacent channels (i.e. $c_p = c_m \pm 1$) and zero otherwise. RTS/CTS messages are activated. The offered load of each traffic source is set to 2,21 Mbps and the packet size to 512 byte, which is the one that saturates our network configuration as it will be showed in the following [20]. Finally, we configured ns-2 to employ the shadowing radio propagation model, in accordance with the interference model formulation (see Section 3.1), with the path loss exponent set to 2.7.

We report only the results for flow F-A, which is the shortest, and flow I-A, which is the longest. The results of flow H-A are very similar, and thus omitted.

Flow F-A

As shown in Fig. 8, presenting the simulation results regarding the traffic flow from node F to node A, the proposed algorithm performs much better than CLICA and CCA. CLICA reaches the same throughput as our solution for link F-C, but it does not provide a good solution for link C-A, which is the most loaded. This is reflected in the end to end delay in Fig. 9, where the proposed solution grants a much smaller delay. In all cases CCA performs poorly. This is due to the fact that all nodes share the same channel subset, therefore increasing the co-channel interference.

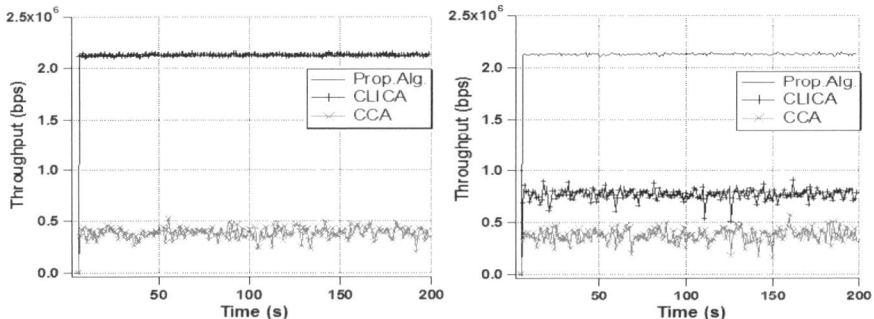

Fig. 8. Throughput in the hop F-C (left) and in the hop C-A (right)

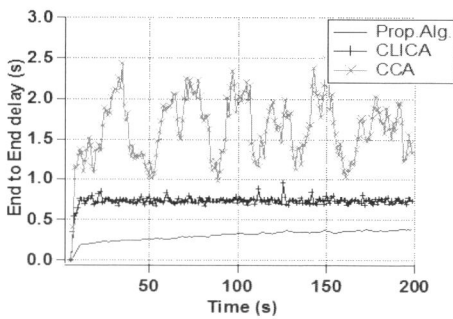

Fig. 9. End to End Delay of flow F-A

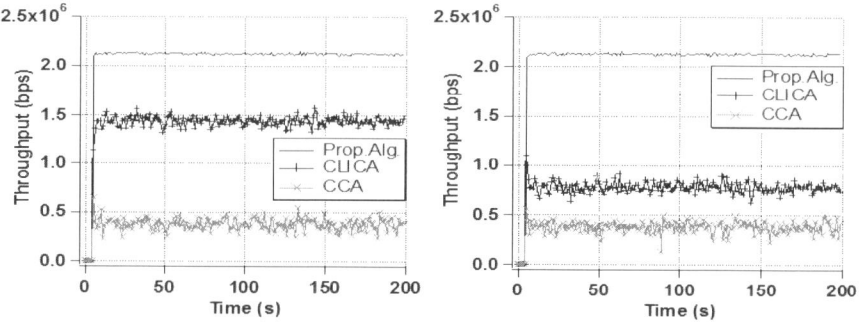

Fig. 10. Throughput in the hop I-L (left) and in the hop L-G (right)

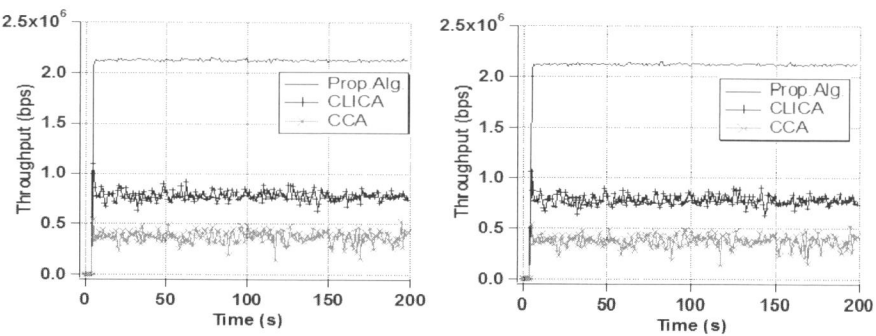

Fig. 11. Throughput in the hop G-D (left) and in the hop D-A (right)

Flow I-A

The simulation results regarding the traffic flow from node *I* to node *A*, which is the longest and more critical, are shown in Fig. 10, and Fig. 11. The figures clearly point out that the proposed algorithm is able to reach the maximum throughput in all hops. This is a very important result because it means that each hop will not experiment

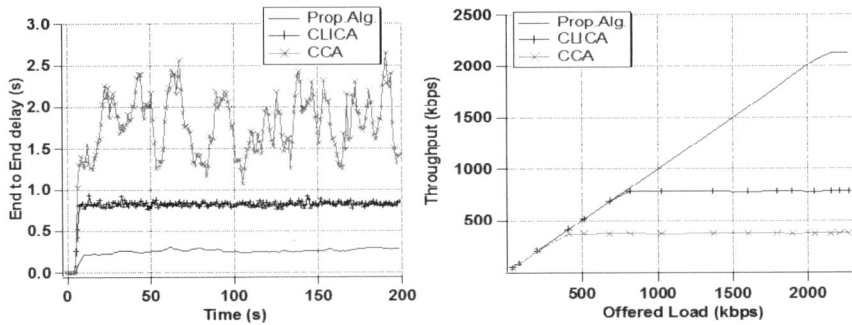

Fig. 12. End to End delay (left) and End to End Throughput (right) of flow I-A

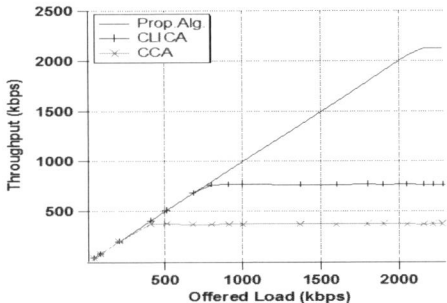

Fig. 13. End to End Throughput flow F-A

inter-flow and intra-flow interference, and therefore each trunk of the flow behaves as if it were isolated. As for the delay, Fig. 12 left confirms the previous results, as our algorithm achieves a manifestly smaller end-to-end delay than CLICA and CCA.

Saturation Point
A very interesting measure of the efficiency of our channel allocation algorithm is given by the saturation point it can reach. We have therefore assessed the end-to-end throughput for different offered loads of the traffic sources. Fig. 12 right and Fig. 13 show the results for flow I-A and flow F-A. As we can see, the proposed solution provides a channel configuration that saturates for an offered load that is much greater than the configurations produced by the other algorithms. While CCA reaches at most 400 kbps, and CLICA arrives at roughly 770 kbps, our solution can achieve up to 2.2 Mbps, which is almost a 300% gain. Also note that this performance gain is available to both flows. This is indeed a noteworthy result that gives a tangible indication on the improvement in network capacity that the proposed solution can provide.

6 Conclusion

The paper presents a channel assignment algorithm for multi-radio WMN that utilizes a very accurate interference model. Beyond the signal power received from other nodes, it also accounts for coupling effects due to the presence of multiple radio modules on the same board. A connectivity graph and a weighted conflict graph are used to formalise the problem. The weights assigned to the edges of the conflict graph are evaluated using both the interference model and the knowledge of the traffic pattern (by means of utilization factors; joint routing and channel assignment is currently under study). This approach revealed to be very convenient. Simulation results showed that the channel configuration produced by our algorithm allows reducing and even removing intra-flow and inter-flow interference. Hence each wireless link can work like an isolated link. As a consequence, the performance in terms of throughput and end-to-end delay is much better than previous solutions (CLICA and CCA). Most remarkably, the channel configuration provided by our algorithm saturates for an offered load that is three times higher than the other channel assignment solutions.

Acknowledgements

The authors wish to thank Lorenzo Bianconi for his useful contribution in the implementation of the described algorithm. This work was supported by the Italian MIUR-FIRB project "Integrated System for Emergency" (InSyEme), under grant number RBIP063BPH.

References

1. IEEE standard 802.11-2007 (June 2007)
2. Ramachandran, K.N., Belding, E.M., Almeroth, K.C., Buddhikot, M.M.: Interference-Aware Channel Assignment in Multi-Radio Wireless Mesh Networks. In: Proc. of IEEE Int. Conf. on Computer Communications (INFOCOM) (April 2006)
3. Ko, B., Misra, V., Padhye, J., Rubenstein, D.: Distributed Channel Assignment in Multi-Radio 802.11 Mesh Networks. In: IEEE Wireless Communications and Networking Conference (WCNC) (March 2007)
4. Raniwala, A., Chiueh, T.-c.: Architecture and algorithms for an IEEE 802.11-based multi-channel wireless mesh network. In: Proc. of IEEE International Conference on Computer Communications (INFOCOM) (March 2005)
5. Marina, M.K., Das, S.R.: A topology control approach for utilizing multiple channels in multi-radio wireless mesh networks. In: Int. Conf. on Broadband Networks (October 2005)
6. Kyasanur, P., Vaidya, N.H.: Routing and interface assignment in multichannel multi-interface wireless networks. In: Proc. of IEEE Wireless Communications and Networking Conference (WCNC) (March 2005)
7. Alicherry, M., Bhatia, R., Li, L.E.: Joint channel assignment and routing for throughput optimization in multi-radio wireless mesh networks. In: Proc. of the ACM Int. Conf. on Mobile Computing and Networking (MOBICOM) (2005)
8. Bahl, P., Chandra, R., Dunagan, J.: SSCH: Slotted Seeded Channel Hopping for Capacity Improvement in IEEE 802.11 Ad-Hoc Wireless Networks. In: Proc. of the ACM Int. Conf. on Mobile Computing and Networking (MOBICOM) (2004)
9. So, J., Vaidya, N.: Multi-channel MAC for ad hoc networks: handling multi-channel hidden terminals using a single transceiver. In: Proc. of the ACM Int. Symposium on Mobile Ad Hoc Networking and Computing (MOBIHOC) (May 2004)

10. Mishra, A., Shrivastava, V., Banerjee, S., Arbaugh, W.: Partially Overlapped Channel Not Considered Harmful. In: Proc. of the joint Int. Conf. on Measurement and Modeling of Computer Systems (June 2006)
11. Cheng, C.-M., Hsiao, P.-H., Kung, H.T., Vlah, D.: Adjacent Channel Interference in Dual-radio 802.11a Nodes and Its Impact on Multi-hop Networking. In: IEEE Global Telecommunications Conference (GLOBECOM) (November 2006)
12. Skalli, H., Ghosh, S., Das, S.K., Lenzini, L., Conti, M.: Channel Assignment Strategies for Multi-radio Wireless Mesh Networks: Issues and Solutions. IEEE Communications Magazine 45(11), 86–95 (2007)
13. Draves, R., Padhye, J., Zill, B.: Routing in Multi-radio, Multi-hop Wireless Mesh Networks. In: Proc. of the ACM Int. Conf. on Mobile Computing and Networking (MOBICOM), Philadelphia, PA (September 2004)
14. IEEE Unapproved draft P802.11s/D1.02 (March 2007)
15. Rosen, K.: Discrete Mathematics and its Applications. McGraw-Hill, New York (1999)
16. Jain, K., Padhye, J., Padmanabhan, V., Qiu, L.: Impact of Interference on Multi-hop Wireless Network Performance. In: Proc. of ACM Int. Conf. on Mobile Computing and Networking (MOBICOM), San Diego, CA (September 2003)
17. Rappaport, T.S.: Wireless Communication System: Principles & Practice, 2nd edn. Prentice Hall, Englewood Cliffs (2002)
18. Raniwala, A., Gopalan, K., Chiueh, T.-C.: Centralized Channel Assignment and Routing Algorithms for Multi-channel Wireless Mesh Networks. ACM SIGMOBILE Mobile Computing and Communications Review (MC2R) (April 2004)
19. Paquereau, L., Helvik, B.E.: A module-based wireless node for ns-2. In: Proc. of the Workshop on NS2: the IP network simulator, Pisa, Italy (2006)
20. Anastasi, G., Borgia, E., Conti, M.: IEEE 802.11b Ad-Hoc Networks: Performance Measurements. Cluster Computing 8(2-3) (July 2005)

Performance Evaluation of a Dynamic Medium Access Control Scheme for Mobile Ad-Hoc Networks

Mamun I. Abu-Tair[1], Geyong Min[1], Qiang Ni[2], and Hong Liu[3]

[1] Department of Computing, School of Informatics, University of Bradford, Bradford, U.K.
{m.i.a.abu-tair,g.min}@brad.ac.uk
[2] Electronic & Computer Engineering Division, School of Engineering & Design,
Brunel University, U.K.
qiang.ni@brunel.ac.uk
[3] Wuhan National Laboratory for Optoelectronics,
Huazhong University of Science and Technology, China
hongliu@hust.edu.cn

Abstract. Recently there have been considerable interests focusing on the enhancement of Mobile Ad-hoc NETworks (MANETs) which are a collection of wireless mobile nodes forming a temporary network without using any centralized access point, infrastructure, or centralized administration. The Distributed Coordination Function (DCF) of IEEE 802.11 Medium Access Control (MAC) protocol has been widely employed in MANETs to manage the shared wireless medium. In DCF, the size of contention window is doubled upon a collision regardless of the network loads. This paper presents a dynamic MAC scheme to improve the performance of MANETs, which applies a threshold of the collision rate to switch between two different functions for increasing the size of contention window and two different mechanisms of resting the size of the contention window based on the status of network loads. The performance of this scheme is investigated and compared to the original DCF using the network simulator NS-2. Moreover, the Random WayPoint (RWP) mobility model is adopted to investigate the effects of mobility on network performance. The performance results reveal that the dynamic scheme is able to achieve the higher throughput and energy efficiency as well as lower end-to-end delay than the original DCF with and without mobility.

1 Introduction

The rapid increase in the number of Personal Digital Assistants (PDA) devices, palmtops and compact laptops has made wireless networks more popular in the network industry. They provide a flexible data communication system that can either replace or extend a wired LAN to provide location independent network access between computation and communication devices using waves rather than a cable infrastructure [4, 10]. Wireless networks are becoming more widely recognized as a general-purpose connectivity alternative for a broad range of business organizations owing to its simplicity, scalability, relative ease of integrating wireless access and

L. Cerdà-Alabern (Ed.): Wireless and Mobility 2008, LNCS 5122, pp. 89–101, 2008.
© Springer-Verlag Berlin Heidelberg 2008

ability for wireless stations to roam throughout the business organizations with the remaining connected to other existing network resources such as servers, printers, and Internet connections.

One of the major types of wireless networks is Mobile Ad-hoc NETworks (MANETs), which consist of a collection of mobile stations communicating with each other without the use of any pre-existent infrastructure. In order to allow mobile stations to share the wireless medium, many practical MANETs have employed the Institute of Electrical and Electronics Engineers (IEEE) 802.11 standards ratified in 1997 that can operate at data rates up to 2 Mbps in the 2.4-GHz Industrial, Scientific and Medical (ISM) band. But the most general business requirements cannot be well supported with this low data rate of the original IEEE 802.11 standard. Recognizing the critical need to support the higher data-transmission rates, the IEEE ratified both 802.11a and 802.11b standards with the rate up to 54 and 11 Mbps in the 5 and 2.4-GHz ISM band, respectively [10, 25]. Moreover, both standards specify the operation of the Medium Access Control (MAC) protocol, which is responsible for control access to the transmission medium with the aim of utilizing the capacity of transmission media in an efficient manner by wireless network devices.

The IEEE 802.11 MAC protocol offers two different methods to support share access to wireless channels; a Distributed Coordination Function (DCF) and an optional Point Coordination Function (PCF) [10]. At the present time, the DCF is the dominant MAC mechanism implemented in the IEEE 802.11-compliant products. The DCF is based on the Carrier Sense Multiple Accesses (CSMA) mechanism, which is a contention-based protocol making certain that the stations first sense the medium before data-transmission. Moreover, the DCF applies a collision avoidance (CA) mechanism which can reduce the probability of collisions using an additional random binary exponential time called back-off time. The main objective of CSMA/CA is to avoid stations transmitting at the same time, which can lead to collisions and corresponding retransmissions [10, 14, 15].

In addition to the common CSMA/CA techniques, the DCF further reduces the possibility of collisions and improves data delivery reliability by adding acknowledgement frames and optional channel reservation frames (i.e., Request-To-Send and Clear-To-Send) to the exchange sequences of its data frames. Different from DCF, the optional coordination function PCF is a centralized scheme designed for infrastructure networks that have a point coordinator operating at the Access Point (AP) to poll and select the next wireless station for data-transmission [10].

The enhancement of the IEEE 802.11 MAC protocol has attracted numerous research efforts [6, 13, 23]. Wu *et al.* [23] have proposed a mechanism to enhance the throughput of DCF by adjusting the scheme of resetting the contention window. Lin and Pan [13] have introduced a Tender back-off algorithm, which adds two more back-off stages to the original Binary Exponential Backoff (BEB) scheme. They have shown that this mechanism can improve the throughput of DCF. However, the authors did not study other QoS performance measures of the proposed mechanism, such as packet delay and loss probability. Chatzimisions *et al.* [6] have proposed a Double Increment Double Decrement (DIDD) back-off algorithm, which decreases the size of contention window half after a successful transmission rather than resetting it to the minimum value. They have conducted an extensive performance study to demonstrate the efficiency of this algorithm.

Multimedia communication over wireless networks has become the driving technology for many important applications, experiencing dramatic market growth and promising revolutionary experiences in personal communication, gaming, entertainment, military, security, environment monitoring, and more. One of the typical wireless multimedia applications is the real-time Variable-Bit-Rate (VBR) video. The results of high-quality measurement traces [2] have indicated that the traffic of compressed VBR videos exhibits fractal-like self-similar and Long Range Dependent (LRD) properties (i.e., traffic burstiness and correlations appearing over many time scales). More recently, the self-similar traffic behaviors have been found in WLANs and MANETs [22].

In order to enhance the performance of the DCF protocol, we propose a new dynamic MAC scheme, which takes the traffic loads of the stations into account. This scheme uses the recent collision rate as a threshold to switch between two different increasing functions (i.e., exponential and quadratic) to deal with transmission collision and employs two different mechanisms to rest the contention window for successful transmission. When the mobile station works under light traffic loads, the contention window increases exponentially in the case of collision and rests to the minimum contention window in the case of successful transmission. However, when the wireless station has heavy traffic loads, the contention window increases quadratically in the case of collision and decreases exponentially in the case of successful transmission. The performance results based on simulation experiments demonstrate that the dynamic MAC scheme outperforms the original DCF in terms of throughput, end-to-end delay and energy efficiency with and without stations mobility. Moreover, the Random WayPoint (RWP) mobility model is adopted to investigate the effects of mobility on network performance.

The rest of this paper is organized as follows: Section 2 introduces DCF MAC protocol of the IEEE 802.11 wireless networks. Section 3 presents the dynamic MAC scheme. Section 4 describes the simulation scenarios and then introduces the self-similar traffic model and the Random WayPoint (RWP) mobility model. Section 5 presents the performance results obtained from simulation experiments. Finally, Section 6 concludes this study.

2   Distributed Coordinated Function (DCF)

The contention-based DCF is the basic access mechanism of IEEE 802.11. Similar to other contention-based MAC protocols, DCF relies on a Carrier Sense Multiple collision Access with Collision Avoidance (CSMA/CA) algorithm to access the shared medium. A station having packets ready for transmission senses whether or not the medium is busy. If it has been idle for longer than the minimum duration called DCF Interference Space (DIFS), the station starts transmission immediately. Otherwise, a back-off time is chosen randomly from the interval $[0, cw]$, where cw represents the contention window [14, 15]. The station starts down-counting its back-off counter by one only if the medium has been detected idle for at least a DIFS. If the medium gets busy due to other transmissions, the back-off counter pauses down-counting and resumes when the medium has been sensed idle for DIFS again [10, 14]. Transmission may proceed when back-off counter has reached zero. Upon detection of a collision, i.e., when the back-off counter of two or more stations reaches zero at the same time, the contention window is doubled according to $cw_i = 2^{k+i-1} - 1$ where i is number of attempts to transmit the frame and k is a constant defining the minimum contention window $cw_{min} = 2^k - 1$ [14]. When the destination station receives frame successfully, it sends an acknowledgment (ACK) frame back to the source station after a Short Inter Frame Space (SIFS) duration. Additionally, to alleviate the hidden station problem, DCF uses optional Request-to-Send/Clear-to-Send (RTS/CTS) frames before packet transmission [10]. This process has been illustrated in Fig. 1.

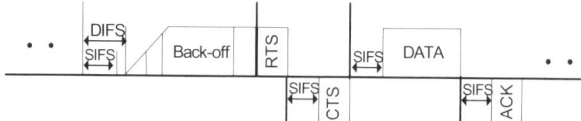

Fig. 1. RTS/CTS of IEEE 802.11 DCF

3   Dynamic MAC Scheme

IEEE 802.11 DCF adopts an exponential back-off algorithm to handle the transmission collision between different wireless stations. However, the DCF does not consider the station situation whether it is under heavy traffic loads or not. When there are few contending stations, the DCF tends to work well in an efficient manner. However, when the number of stations increases, the slow increase in the case of transmission collision and the rapid decrease of contention window under successful transmission can lead to significant performance degradation. To handle this problem, this paper presents a dynamic MAC scheme to manage wireless medium. This scheme uses the recent collision rate as a threshold to switch between two different increasing functions (i.e., exponential and quadratic) in the case of transmission collision and two different decreasing mechanisms in the case of successful transmission. More specifically, when the wireless station works under the light traffic loads (i.e. the collision rate is low), this scheme operates exactly as DCF: (1) increases the contention window exponentially in the case of collision; (2) reduces the contention window to minimum contention window in the case of successful transmission. However, when the wireless station has the heavy traffic loads and the collision rate is high, the dynamic scheme uses: (1) the quadratic increasing function to increase the contention window size in the case of collision; (2) the exponential decreasing function to reduce the contention window size in the case of successful transmission. This scheme can avoid the increasing number of collisions caused by unsuitable small contention window sizes or by resting the contention window to the minimum contention window size after any successful transmission.

Similar to [9, 16, 20, 24], the recent collision rate can be calculated by dividing the time domain of the network connection into continuous intervals with the specific number of slot times. At the end of each interval, the mobile station computes the collision rate as

$$\chi_{curr}^{j} = \frac{(num\_collisions)^{j}}{(num\_packets\_sent)^{j}} \tag{1}$$

Where $(num\_collisions)^{j}$ and $(num\_packets\_sent)^{j}$ are the number of collisions and the number of packets successfully sent at the jth interval, respectively.

Furthermore, in order to precisely represent the long-term and short-term network conditions, an Exponentially Weighted Moving Average (EWMA) method is used as a smoother for the estimated value of the collision rate. In particular, let χ_{ave}^{j} denote the average collision rate after the jth interval period. χ_{ave}^{j} is given by

$$\chi_{ave}^{j} = (1-\delta) * \chi_{curr}^{j} + \delta * \chi_{ave}^{j-1} \quad \text{where } 0 < \delta < 1 \tag{2}$$

Figs. 2-4 show the flow chart of the main procedures used in the dynamic MAC scheme. The decrement, updating and increment procedures are triggered after successful transmission, at the end of the jth time interval and after transmission collision, respectively.

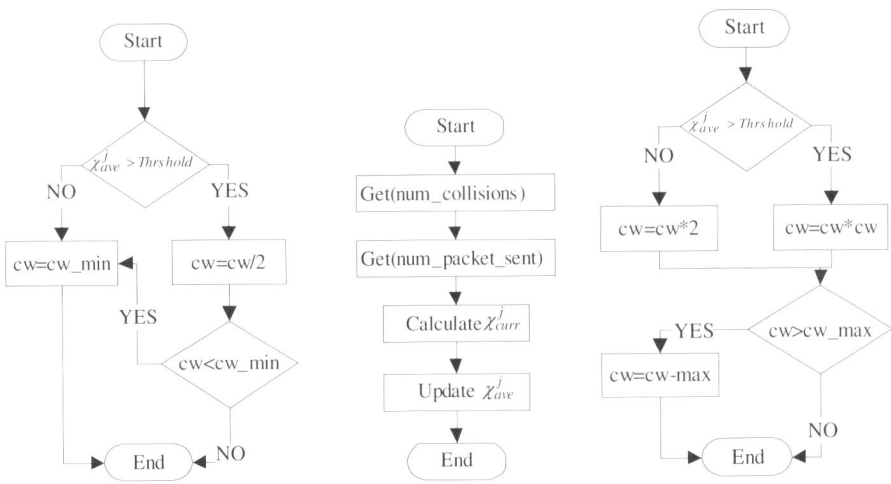

Fig. 2. Decrement procedure   **Fig. 3.** Updating procedure   **Fig. 4.** Increment procedure

4 Simulation Scenarios and Configuration

The well-known network simulator NS-2 [17] has been adopted to conduct our simulation experiments. This section will present the experiment scenarios and how to configure simulation parameters. The simulation scenarios studied in this research was designed to investigate the performance of the dynamic MAC scheme in MANETs.

The scenarios are composed of up to 50 mobile stations and a common mobile station serving as a sink for all traffic. Each mobile station operates under the IEEE 802.11a standard at a data rate of 24Mbps [14]; the setting of physical layer (PHY) and MAC parameters are shown in Table 1. All stations are used RTS/CTS mechanism and Destination-Sequenced Distance-Vector (DSDV) routing protocols [19].

4.1 Traffic Model

Appropriate models that can accurately capture the properties of real-world network traffic are required for effective and reliable performance evaluation of the MANET protocol. Recent measurement studies have indicated that traffic in the compressed Variable-Bit-Rate (VBR) video and MANETs exhibits self-similar nature. According to the measurement results of Star Wars video trace [2, 7], each station generates real-time video traffic with rate of 360 Kbit/s and Hurst parameter of 0.7. Hurst parameter is used to characterize the degree of traffic self-similarity. Such traffic can be generated by the superposition of many ON/OFF sources in which the ON and OFF periods follow Pareto distributions (i.e., with high variability or infinite variance) [16]. The lengths of the ON and OFF periods are determined by $K(x^{-1/a} - 1)$ where x is a uniformly distributed random number between 0 and $1, a = 3 - 2H$, $K = m(a - 1)$ and m is the mean length of ON and OFF period, respectively [11]. In the simulation experiments, we used the superposition of five Pareto-distributed ON/OFF flows with the mean ON period of 10ms and OFF period of 100ms to generate video traffic.

Table 1. NS-2 PHY Parameters for IEEE 802.11a

SlotTime	9 µs
CCATime	3 µs
RxTxTurnaroundTime	2 µs
SIFSTime	16 µs
PreambleLength	96 bits
PLCPHeaderLength	40 bits
PLCPDataRate	6 Mbps
cw_{min}	15
cw_{max}	1023
δ	0.8
Collision Threshold	0.5

4.2 Mobility Model

The Random WayPoint (RWP) mobility model proposed by Johnson and Maltz [4] is the most popular mobility model used for performance analysis of MANETs. Two key parameters of the RWP models are $V_{\max}$ and T_{pause} where $V_{\max}$ is the maximum velocity for every mobile station and T_{pause} is the pause time. A mobile station in the RWP model selects a random destination and a random speed between [0, $V_{\max}$], and then moves to the selected destination at the selected speed. Upon reaching the destination, the mobile station stops for some pause time T_{pause}, and then repeats the process by selecting a new destination and speed and resuming the movement. Fig. 5 shows a movement trace of a mobile station using a RWP mobility model. The mobility scenario generation and analysis tool, BonnMotion [3], is used to generate the mobility scenarios for simulation experiments in this study.

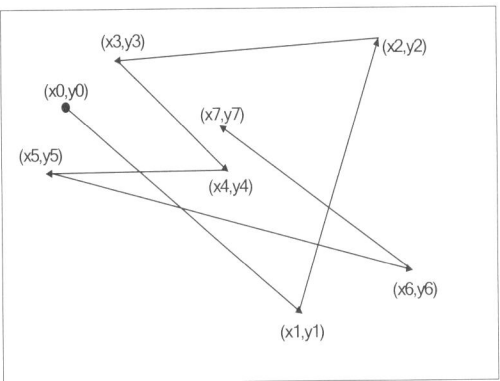

Fig. 5. Example of mobile station's movement according to the RWP mobility model

5   Analysis and Evaluation of Various Performance Metrics

The simulation experiments aim to investigate the performance of the dynamic MAC scheme in the MANETs in terms of throughput, end-to-end delay and energy efficiency rate in the presence of multimedia traffic and stations mobility. The average throughput is calculated as the amount of data actually delivered to the destination during each time unit. Many factors affect the throughput, including the efficiency of collision avoidance, medium utilization, latency, and control overhead. The end-to-end delay is defined as the time elapsed between the packet originate from the source station and the successful arriving to the destination station. We measure the end-to-end delay to find out how well the dynamic MAC scheme accommodates the multimedia services. The energy efficiency rate is referred to the number of data delivered per joule energy consumed.

5.1   Throughput

As an effort to investigate the performance of the Dynamic MAC scheme, Figs. 6 and 7 compare the throughput of the original and dynamic DCF versus the number of

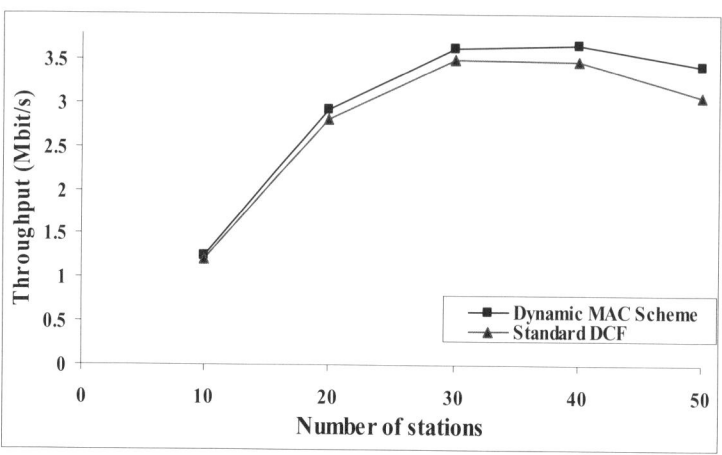

Fig. 6. Throughput comparison between the dynamic MAC scheme and the original DCF for the static scenario

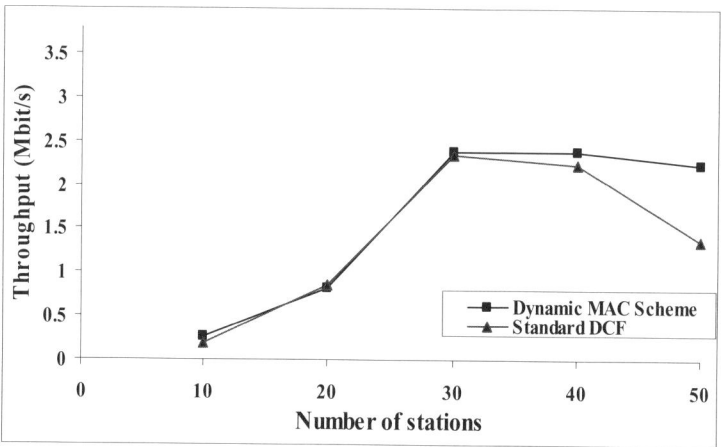

Fig. 7. Throughput comparison between the dynamic MAC scheme and the original DCF for the RWP mobility model scenario

mobile stations. For the static scenario (Fig. 6), the dynamic MAC scheme achieves a higher throughput than the original DCF in all cases. For example when there are 50 active stations, the throughput of the dynamic scheme and the original DCF is around 3.4 Mbit/s and 3.0 Mbit/s, respectively.

It is worth noting that setting of a small size of contention window and slow increasing of contention window in the case of collision used by the original DCF could degrade the system throughput when there are a large number of contenting stations. Moreover, Fig. 7 shows that the dynamic scheme achieves the higher throughput than the original DCF under RWP mobility model. For example when there are 50 active stations, the throughput of the dynamic scheme and the DCF is around 2.24 Mbit/s and 1.36 Mbit/s, respectively. Additionally, it is clearly observed that the movement of the mobile stations has a deteriorating effect on the throughput of the MANETs. As a result, the throughput of the dynamic MAC scheme increases up to 12%, 37% under static and RWP mobility model, respectively, compared to the original DCF.

5.2 End-to-End Delay

Figs. 8 and 9 represent the comparison of the mean end-to-end delay between the original DCF and dynamic MAC scheme. As shown in the figures, the end-to-end delay of the dynamic MAC scheme is lower than that of the original DCF in most cases under both static and RWP mobility model. This is because the dynamic MAC scheme adopts the quadratic increasing and exponential decreasing functions in the case of collision and successful transmission, respectively, which provides the contenting station with the large contention windows. This mechanism can overcome the increasing delay caused by the small contention window, which increases the collision probability. In general, the dynamic scheme maintains around 20% lower of the mean end-to-end delay than the original DCF for most cases. Moreover, using two different increasing functions and two different resting mechanisms to the contention window in the dynamic scheme can provide the mobile stations with a proper contention window size adaptable to the network status (a large contention window when there are more contenting stations and a small one when there are few contenting stations). The use of this technique directly affects the end-to-end delay. As a result, the dynamic MAC scheme has comparable end-to-end delay under the light traffic loads and lower end-to-end delay under the heavy loads compared to the original DCF in most cases.

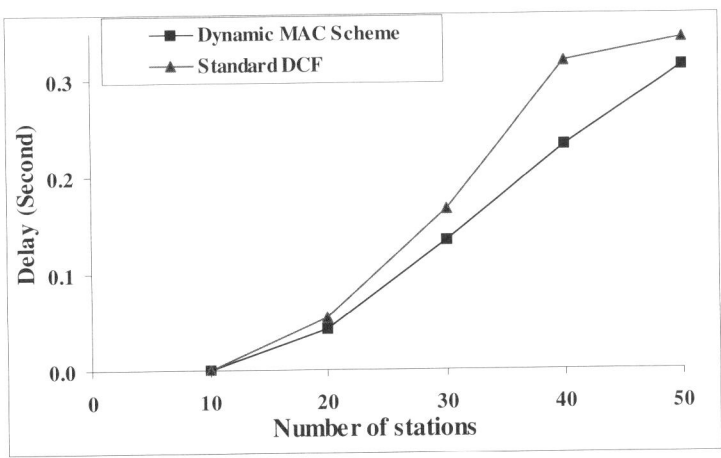

Fig. 8. End-to-end delay comparison between the dynamic MAC scheme and the original DCF for the static scenario

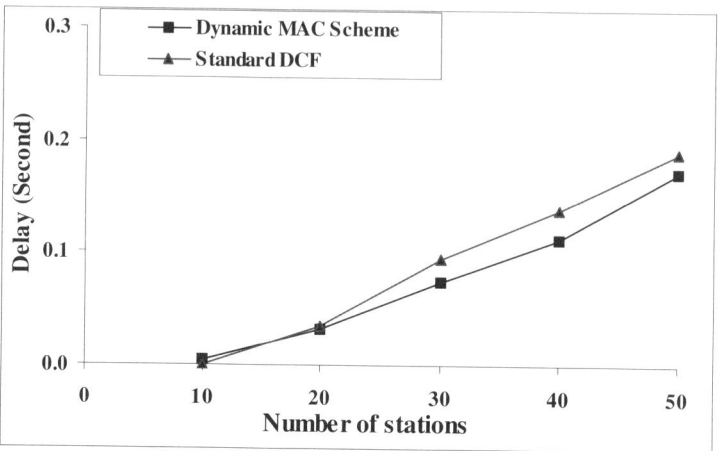

Fig. 9. End-to-end delay comparison between the dynamic MAC scheme and the original DCF for the RWP mobility model scenario

5.3 Energy Efficiency

In addition to throughput and end-to-end delay this study takes the energy efficiency into consideration. Figs. 10 and 11 show the comparison of the energy efficiency rate between the dynamic MAC scheme and the Original DCF under both static and RWP mobility model. As shown in the figures, the dynamic MAC scheme is more energy-efficient than the original DCF in most cases under both scenarios. This is because, compared to the original DCF, the dynamic scheme achieves the few number of collisions which increases the number of successful transmissions and saves the energy of the mobile station. More specifically, the rate of the energy efficiency of the

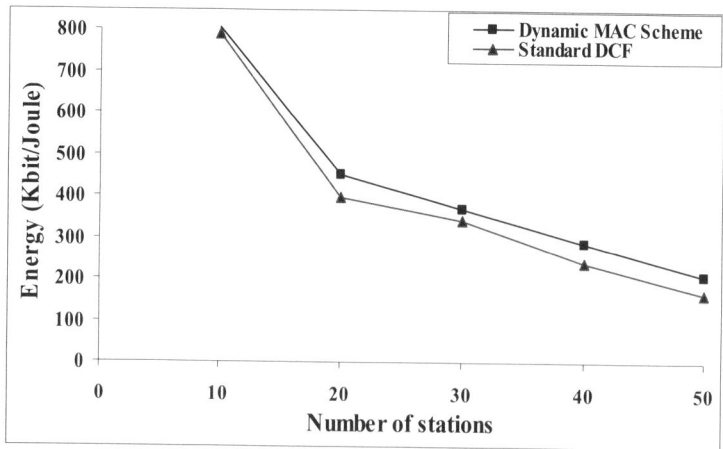

Fig. 10. Energy efficiency rate comparison between the dynamic MAC scheme and the original DCF for the static scenario

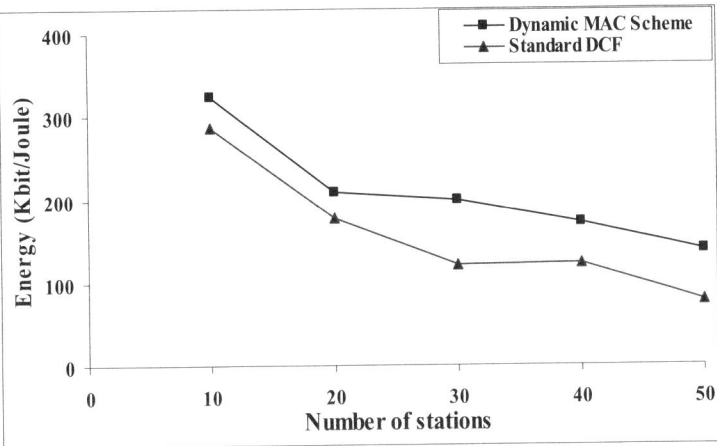

Fig. 11. Energy efficiency rate comparison between the dynamic MAC scheme and the original DCF for the RWP mobility model scenario

dynamic MAC scheme increases up to 21% and 39% for static and RWP mobility models, respectively, compared to the original DCF.

6 Conclusions

Mobile Ad hoc NETworks (MANETs) play an important role in civilian forums such as campus recreation, conferences, and electronic classrooms *etc*. Many MAC protocols have been reported to mange and control the shared wireless medium. This paper has presented a dynamic MAC scheme in order to enhance the performance of multimedia services in the IEEE 802.11 MANETs. This scheme employs the collision rate as a threshold to (1) switch between an exponential and quadratic increasing function to increase the contention window in the case of collision and (2) reduce the contention window to minimum contention window or decrease it exponentially in the case of successful transmission. Performance evaluation of this dynamic scheme has been conducted using the network simulator NS-2 under static and Random WayPoint (RWP) mobility model. The performance results have shown that the dynamic scheme outperforms the original DCF in terms of throughput, end-to-end delay and energy efficiency rate with and without mobility. More specifically, for the static scenario the throughput of the dynamic scheme increases up to 12%, the energy efficiency rate increases 21% and the end-to-end delay decreases 36%. Additionally, for the RWP mobility model scenario the throughput of the dynamic scheme increases up to 37%, the energy efficiency rate increases 39% and the end-to-end delay decreases 25%.

Acknowledgements. This work is supported in part by the UK EPSRC Grant EP/C525027/1 and in part by the EC NoE Euro-FGI (NoE 028022).

References

1. Aad, I., Castelluccia, C.: Differentiation mechanisms for IEEE 802.11. In: Proc.20th Annual Joint Conference of the IEEE Computer and Communications Societies, IEEE (INFOCOM 2001), vol. 1, pp. 209–218 (2001)
2. Beran, J., Sherman, R., Taqqu, M.S., Willinger, W.: Long-range dependence in variable-bit-rate video traffic. IEEE Trans. Communications 43, 1566–1579 (1995)
3. BonnMotion, A mobility scenario generation and analysis tool, http://www.cs.uni-bonn.de/IV/BonnMotion/
4. Broch, J., Maltz, D.A., Johnson, D.B., Hu, Y.-C., Jetcheva, J.: A performance comparison of multi-hop wireless ad hoc network routing protocols. In: Proc. of the Fourth Annual ACM/IEEE International Conference on Mobile Computing and Networking (Mobicom 1998). ACM, New York (1998)
5. Camp, T., Boleng, J., Davies, V.: A Survey of Mobility Models for Ad Hoc Network Research. Wireless Communication and Mobile Computing (WCMC): Special issue on Mobile Ad Hoc Networking: Research, Trends and Applications 2(5), 483–502 (2002)
6. Chatzimisions, P., Vitasas, V., Boucouvalas, A., Tsoulfa, M.: Achieving performance enhancement in IEEE 802.11 WLANs by using DIDD backoff mechanism. International Journal of Communication Systems 20(1), 23–41 (2007)
7. Chen, B., Lucic, Z., Trajkovic, L.: Simulation and wavelet analysis of packet traffic (2005), http://www.ensc.sfu.ca/people/faculty/ljilja/cnl/pdf/bruce_asi2002.pdf
8. Chuah, C., Katz, R.: Characterizing packet audio streams from Internet multimedia applications. In: Proc. IEEE International Conference on Communication (ICC 2002), vol. 2, pp. 1199–1203 (2002)
9. Gannoune, L.: A non-linear dynamic tuning of the minimum contention window (CWmin) for enhanced service differentiation in IEEE 802.11 ad-hoc networks. In: Proc. Vehicular Technology Conference (VTC 2006), vol. 3, pp. 1266–1271 (Spring) (2006)
10. Gast, M.: 802.11 wireless networks: the definitive guide, vol. 1. O'Reilly, Sebastopol (2002)
11. Hassan, M., Jain, R.: High performance TCP/IP networking concepts, issues and solutions. Pearson Prentice Hall, London (2004)
12. Hong, X., Gerla, M., Pei, G., Chiang, C.-C.: A group mobility model for ad hoc wireless networks. In: Proc. ACM International Workshop on Modeling, Analysis, and Simulation of Wireless and Mobile Systems (MSWiM) (1999)
13. Lin, I.-H., Pan, J.-Y.: Throughput analysis of a novel backoff algorithm for IEEE 802.11 WLANs. In: Proc. Wireless Telecommunications Symposium, pp. 85–90 (2005)
14. Mangold, S., Choi, S., Hiertz, G., Klein, O., Walke, B.: Analysis of IEEE 802.11e for QoS support in wireless LANs. IEEE Wireless Communications 10(6), 40–50 (2003)
15. Mangold, S., Choi, S., May, P., Klein, O., Hiertz, G., Stibor, L.: IEEE 802.11e wireless LAN for quality of service. In: Proc. European Wireless Conference, Italy (2002)
16. Ni, Q.: Performance analysis and enhancements for IEEE 802.11e wireless networks. IEEE Network 19(4), 21–27 (2005)
17. Ns-2 simulator, http://www.isi.edu/nsnam
18. Paxson, V., Floyd, S.: Wide-area traffic: the failure of Poisson modeling. IEEE/ACM transaction on networking 3(3), 226–244 (1995)
19. Perkins, C.E., Bhagwat, P.: Highly dynamic destination-sequenced distance-vector routing (DSDV) for mobile computers. ACM SIGCOMM Computer Communication Review 24(4), 234–244 (1994)

20. Romdhani, L., Ni, Q., Turletti, T.: Adaptive EDCF: enhanced service differentiation for IEEE 802.11 wireless ad hoc networks. In: Proc. IEEE Wireless Communications and Networking Conference (WCNC 2003), New Orleans, Louisiana, USA (2003)
21. Tao, Z., Panwar, S.: Throughput and delay analysis for the IEEE 802.11e enhanced distributed channel access. IEEE Trans. Communications 54(4), 596–603 (2006)
22. Tickoo, O., Sikdar, B.: On the impact of IEEE 802.11 MAC on traffic characteristics. IEEE Journal on Selected Areas in Communications 21(2), 189–203 (2003)
23. Wu, H., Cheng, S., Peng, Y., Long, K., Ma, J.: IEEE 802.11 distributed coordination function (DCF): analysis and enhancement. In: Proc. IEEE International Conference (ICC 2002), vol. 1, pp. 605–609 (2002)
24. Zhu, H., Cao, G., Yener, A., Mathias, A.: EDCF-DM: a novel enhanced distributed coordination function for wireless ad hoc networks. In: Proc. IEEE International Conference on Communication (ICC 2004), vol. 7, pp. 3886–3890 (2004)
25. IEEE WG 802.11, Part 11: Wireless LAN Medium Access Control (MAC) and Physical Layer (PHY) Specifications, ISO/IEC 8802-11:1999(E), IEEE Standards 802.11 (1999)

Teletraffic Performance Analysis of Multi-class OFDM-TDMA Systems with AMC

Hua Wang and Villy B. Iversen

Department of Communications, Optics & Materials
Technical University of Denmark, Lyngby, Denmark
{huw,vbi}@com.dtu.dk
http://www.com.dtu.dk

Abstract. In traditional channelized multiple access systems, such as TDMA and FDMA, each user is assigned a fixed amount of bandwidth during the whole connection time, and the teletraffic performance in terms of time congestion, call congestion and traffic congestion can easily be obtained by using the classical Erlang-B formula. However, with the introduction of adaptive modulation and coding (AMC) scheme employed at the physical layer, the allocation of bandwidth to each user is no longer deterministic, but dynamically based on the wireless channel conditions. Thus a new connection attempt will be blocked with a certain probability depending on the state of the system and the bandwidth requirement of the connection attempt. In this paper, we present an integrated analytical model of multi-rate loss system with state-dependent blocking to evaluate the performance of multi-class OFDM-TDMA systems with AMC scheme.

Keywords: OFDM, AMC, performance evaluation, state-dependent blocking.

1 Introduction

Future mobile communication systems will provide not only speech and low-speed data services, but also high-speed data services such as wireless multimedia applications ranging from kilobits to megabits per second. This can be achieved by operating the air interface with Orthogonal Frequency Division Multiplexing (OFDM), which is immune to intersymbol interference and frequency selective fading, as it divides the frequency band into a group of mutually orthogonal subcarriers, each having a much lower bandwidth than the coherence bandwidth of the channel. Recently, OFDM-based systems have become a popular choice for such an endeavor. The IEEE 802.16 standard, for instance, has adopted OFDM-TDMA and OFDMA as two transmission schemes at the 2-11 GHz band.

The economical usefulness of a system is effectively measured by the Erlang capacity, which is generally defined as the maximum traffic load the system can support when the blocking probabilities at the call admission control (CAC)

L. Cerdà-Alabern (Ed.): Wireless and Mobility 2008, LNCS 5122, pp. 102–112, 2008.

level do not exceed a certain thresholds. Many models have been proposed at separate layers, e.g., Rayleigh, Rician and Nakagami fading models at the physical layer [8], and queuing models at the data link layer [6]. In traditional channelized multiple access systems, e.g., TDMA and FDMA, each user is assigned a fixed amount of bandwidth during the whole connection time, and the Erlang capacity can easily be obtained by using the well-known Erlang-B formula. One important assumption for applying the classical Erlang-B formula is that the capacity of each channel is constant and the capacity assignment for each connection is fixed. This is true in wired networks such as the traditional public switched telephone network (PSTN). However, in wireless networks, the channel capacity of a wireless link is time-varying due to multipath fading and Doppler shift. Thus the Erlang-B formula cannot directly be used to calculate the blocking probability. Furthermore, unlike wired networks, even if large bandwidth is allocated to a certain wireless connection, the QoS requirements may not be satisfied when the channel experiences deep fades.

In order to enhance the spectrum efficiency while maintaining a target packet error rate (PER) over wireless links, adaptive modulation and coding (AMC) scheme has been widely adopted to match the transmission rate to time-varying channel conditions. With AMC, the allocation of bandwidth to each user is no longer deterministic (i.e., a fixed amount of bandwidth), but depends on the channel conditions in a dynamic way. Outage is defined to occur when the total number of time slots required by the admitted users exceeds the total number of available time slots. Therefore, a new connection may be blocked with a certain probability depending on the state of the system and the bandwidth requirement of this new connection.

An analytical model to investigate the performance of transmissions over wireless links is developed in [1], where a finite-length queuing is coupled with AMC. However, the authors only concentrate on a single-user case. Reference [2] calculates the Erlang capacity of WiMAX systems with fixed modulation scheme, where two traffic classes, streaming and elastic flows, are considered. In reference [3], the authors evaluate the Erlang capacity of a multi-class TDMA system with AMC by separating the calculation of blocking and outage probabilities. Performance analysis of OFDM systems has so far been conducted primarily by simulations. An analytical framework to evaluate the teletraffic performance in terms of time congestion, call congestion and traffic congestion of multi-user multi-class OFDM-TDMA systems with AMC scheme is still missing. In this paper to achieve this goal, we propose an integrated analytical model of multi-rate loss system with state-dependent blocking.

The rest of the paper is organized as follows. In Section 2, we introduce the system model, which includes OFDM transmission with AMC and calculation of state dependent blocking probabilities. In Section 3, an analytical model of multi-rate loss system with state-dependent blocking is presented with relevant performance measures. Numerical results are given in Section 4. Finally, conclusions are drawn in Section 5.

2   System Model

We consider an infrastructure-based wireless access network, where connections are established between a base station (BS) and mobile stations (MSs). Several service classes with different data rate requirements are supported in the system. Users from each service class arrive at the cell in a random order. The call admission control (CAC) module decides whether an incoming call should be admitted or not, based on the current state of the system and the bandwidth requirement of the call. We assume that the BS has perfect knowledge of the channel state information (CSI) of each subchannel of each connection. We further assume that each subchannel is frequency flat and that the channel quality remains constant within a frame, but may vary from frame to frame.

2.1   OFDM Transmission with AMC

We consider an OFDM-TDMA system with M subchannels. At the physical layer, the time axis is divided into frames. A frame is further divided into K time slots, each of which may contain one or more OFDM symbols. Users transmit in the assigned time slots over all subchannels. Adaptive modulation and coding scheme is employed to adjust the transmission mode in each subchannel dynamically according to the time-varying channel conditions.

We assume that each subchannel follows a Rayleigh fading. For flat Rayleigh fading channels, the received SNR on subchannel m is a random variable γ_m with probability density function (pdf) [1]:

$$p_\gamma(\gamma_m) = \frac{1}{\overline{\gamma}_m} \exp\left(-\frac{\gamma_m}{\overline{\gamma}_m}\right) \tag{1}$$

where $\overline{\gamma}_m$ is the average SNR over subchannel m.

The design objective of AMC is to maximize the data rate by adjusting the transmission parameters according to channel conditions, while maintaining a prescribed packet error rate (PER) P_0. Let N denote the total number of transmission modes available (e.g., $N = 5$). Assuming constant power transmission, we divide the entire SNR range into $N + 1$ non-overlapping consecutive intervals with boundaries denoted as $\{\Gamma_n\}_{n=1}^{N+1}$. Specifically, mode n is chosen when $\gamma_m \in [\Gamma_n, \Gamma_{n+1})$. Therefore, with Rayleigh fading, transmission mode n will be chosen on subchannel m with probability:

$$P_m(n) = \int_{\Gamma_n}^{\Gamma_{n+1}} p_\gamma(\gamma_m)\, d\gamma_m = \exp\left(-\frac{\Gamma_n}{\overline{\gamma}_m}\right) - \exp\left(-\frac{\Gamma_{n+1}}{\overline{\gamma}_m}\right) \tag{2}$$

Let $\overline{PER}_{m,n}$ denote the average PER corresponding to mode n on subchannel m. It can be obtained in closed-form as [1]:

$$\overline{PER}_{m,n} = \frac{1}{P_m(n)} \int_{\Gamma_n}^{\Gamma_{n+1}} \alpha_n \exp(-g_n\gamma)\, p_\gamma(\gamma)\, d\gamma \tag{3}$$

where α_n, g_n are the mode dependent parameters shown in Table 1. The average PER of AMC can then be computed as the ratio of the average number of packets in error over the total average number of transmitted packets:

$$\overline{\text{PER}} = \frac{\sum_{m=1}^{M} \sum_{n=1}^{N} R_n \, P_m(n) \, \overline{\text{PER}}_{m,n}}{\sum_{m=1}^{M} \sum_{n=1}^{N} R_n \, P_m(n)} \tag{4}$$

where R_n is the number of bits carried per symbol in transmission mode n as shown in Table 1.

The algorithm for determining the thresholds $\{\Gamma_n\}_{n=1}^{N+1}$ with the prescribed $\overline{\text{PER}} = P_0$ is described in details in [1].

Table 1. Transmission modes with convolutionally coded modulation [1]

	Mode 1	Mode 2	Mode 3	Mode 4	Mode 5
Modulation	BPSK	QPSK	QPSK	16QAM	64QAM
Coding rate	1/2	1/2	3/4	3/4	3/4
R_n (bits/sym)	0.5	1.0	1.5	3.0	4.5
α_n	274.7229	90.2512	67.6181	53.3987	35.3508
g_n	7.9932	3.4998	1.6883	0.3756	0.0900

Let $\mathcal{R}_m$ be a random variable with probability mass function f_m, denoting the number of bits that can be transmitted over subchannel m in one time slot.

$$\mathcal{R}_m \in \{sR_0, sR_1, \cdots, sR_N\}$$
$$f_m(sR_n) = \mathbb{P}(\mathcal{R}_m = sR_n) = P_m(n) \tag{5}$$

where s is the number of symbols per time slot.

Let random variable $\mathcal{R} = \sum_{m=1}^{M} \mathcal{R}_m$ denote the number of bits that can be transmitted over all subchannels in one time slot. Based on the assumption that the channel quality of each subchannel is independent identically distributed (i.i.d.), the probability mass function (pmf) of $\mathcal{R}$, denoted as $f_{\mathcal{R}}$, can be obtained by convolving the pmf of each subchannel f_m as follows:

$$f_{\mathcal{R}} = f_1 \otimes f_2 \cdots \otimes f_M \tag{6}$$

where $a \otimes b$ denotes discrete convolution.

2.2 State Dependent Blocking Probability

Assume the system accommodates L types of service classes, each of which requires a constant bit rate of r_i bits per frame. In multi-class systems, different service classes with different bit rate requirements need different channel bandwidth in terms of time slots. Thus it would be beneficial for teletraffic calculations if we could specify a common channel bandwidth which we may call a *unit channel*. The higher the required accuracy (i.e., bandwidth granularity), the smaller a unit channel we have to specify. Let us define r_{unit} be the constant bit rate of a unit channel and let $d_i = r_i/r_{\text{unit}}$ be number of unit channels needed to establish one connection of service class i. Due to the time-varying nature of wireless channels, the number of time slots occupied by a unit channel can be modeled by a random variable $\mathcal{D}_{\text{unit}} = r_{\text{unit}}/\mathcal{R}$, with probability mass function $f_{\mathcal{D}_{\text{unit}}}$, which can be easily obtained from $f_{\mathcal{R}}$.

In AMC scheme, the modulation and coding rate is chosen according to time-varying channel conditions. As a consequence, the number of time slots allocated to each user is varying on a frame by frame basis. Outage is defined to occur when the total number of time slots required by the admitted users exceeds the total number of available time slots. If outage occurs, the admitted users may not get the prescribed data rate and thus the QoS will be degraded. Suppose that the system is in state x (x unit channels are currently occupied by the admitted users), then the total number of time slots required by the x unit channels can be modeled by the sum of x i.i.d. random variables $\mathcal{D}_x = \sum_1^x \mathcal{D}_{\text{unit}}$ with probability mass function $f_{\mathcal{D}_x}$:

$$f_{\mathcal{D}_x} = \underbrace{f_{\mathcal{D}_{\text{unit}}} \otimes f_{\mathcal{D}_{\text{unit}}} \cdots \otimes f_{\mathcal{D}_{\text{unit}}}}_{x \text{ times}} \tag{7}$$

and the outage probability in state x can be calculated as:

$$
\begin{aligned}
\mathrm{P}_{\text{outage}}(x) &= \mathbb{P}(\mathcal{D}_x > K) \\
&= 1 - \mathbb{P}(\mathcal{D}_x \leq K) \\
&= 1 - \sum_{i \leq K} f_{\mathcal{D}_x}(i)
\end{aligned}
\tag{8}
$$

When a new connection arrives at the system, it will be accepted with a certain probability under the condition that the acceptance of the new connection will keep the system outage probability below a predefined threshold. Therefore, the blocking probability of a new connection is a random variable depending on the state of the system and the bandwidth requirement of the new connection. Specifically, assume that a new single-channel call arrives at the time instance when the system is in state x. The call admission control module first checks the outage probability of current state $\mathrm{P}_{\text{outage}}(x)$. If it is larger than the threshold, the new call is always rejected. If it is smaller than the threshold, the CAC module estimates the outage probability of the next state $\mathrm{P}_{\text{outage}}(x+1)$. If it is smaller than the threshold, the new call is accepted with probability one, otherwise, the new call is accepted with probability $p \in (0,1)$. The value of p is determined under the condition that the estimated system outage probability $\mathrm{P}_{\text{outage}}(x+p)$ is below the threshold. For example, if a new single-channel call arriving in state x is accepted with probability p, the estimated outage probability can be calculated as: $\mathrm{P}_{\text{outage}}(x+p) = 1 - \sum_{i \leq K} f_{\mathcal{D}_{x+p}}(i)$, where $\mathcal{D}_{x+p} = \mathcal{D}_x + \mathcal{D}_p = \sum_1^x \mathcal{D}_{\text{unit}} + p \cdot \mathcal{D}_{\text{unit}}$ is a random variable with probability mass function $f_{\mathcal{D}_{x+p}} = f_{\mathcal{D}_x} \otimes f_{\mathcal{D}_p}$, where $f_{\mathcal{D}_p}$ can be easily obtained from $f_{\mathcal{D}_{\text{unit}}}$.

Let us define the blocking probability of a new single-channel call in state x as $b_x = 1 - a_x$, where a_x is the acceptance probability in state x, determined as follows:

$$
a_x = \begin{cases}
1 & \text{if } \mathrm{P_{outage}}(x+1) \leq \mathrm{Out_{Th}} \\
0 & \text{if } \mathrm{P_{outage}}(x) > \mathrm{Out_{Th}} \\
\max \left\{ p : \mathrm{P_{outage}}(x+p) \leq \mathrm{Out_{Th}} \right\} & \text{else}
\end{cases} \tag{9}
$$

where $\mathrm{Out_{Th}}$ is the predefined outage probability threshold.

For a d-channel call, we have to choose the acceptance probability in state x as a function of the number of unit channels currently occupied and the bandwidth request as follows [5]:

$$
a_{x,d} = 1 - b_{x,d} = \prod_{j=x}^{x+d-1} (1 - b_j) \tag{10}
$$
$$
= (1 - b_x)(1 - b_{x+1}) \cdots (1 - b_{x+d-1})
$$

Notice that $b_x = b_{x,1}$. This corresponds to that a d-channel call chooses one unit channel d times, and it is accepted only if all d unit channels are successfully obtained. This is a quite natural requirement as we assume full accessibility. In the next section, we will see that this is a necessary and sufficient condition for maintaining the reversibility of the process.

3   Analytical Model

We evaluate the performance of the above mentioned multi-class OFDM-TDMA systems with AMC scheme by using the classical teletraffic model of multi-rate loss system with state-dependent blocking.

3.1   Traffic Model

We use the *BPP* (Binomial, Poisson & Pascal) traffic model in our analysis [6]. This model is insensitive to the service time distributions, thus it is very robust for applications. Each traffic stream i is characterized by the offered traffic A_i, the peakedness Z_i and the number of unit channels d_i needed for establishing one connection. The offered traffic A_i is usually defined as the average number of connection attempts per mean holding time. Peakedness Z_i is the variance/mean ratio of the state probabilities of a traffic stream when the system capacity is infinite, and it characterizes the arrival process. For $Z_i = 1$, we have a Poisson arrival process, whereas for $Z_i < 1$, we have a finite number of users and more smooth traffic (Engset case). Engset traffic can alternatively be characterized by the number of traffic sources S and the offered traffic per idle source β. We have the following relations between the two presentations [4]:

$$
A = S \cdot \frac{\beta}{1 + \beta} \qquad Z = \frac{1}{1 + \beta}
$$
$$
\beta = \frac{1 - Z}{Z} \qquad S = \frac{A}{1 - Z} \tag{11}
$$

For $Z_i > 1$, the model corresponds to a more bursty arrival process, called Pascal traffic.

3.2   Algorithms for Calculating Global State Probabilities

The call-level characteristics of multi-class OFDM-TDMA systems with AMC described in Section 2 can be modeled by a multi-dimensional Continuous Time Markov Chain (CTMC). As an example, we consider a system supporting two traffic streams with different data rates. The arrival process of both streams follows a Poisson process with rate λ_1 and λ_2 respectively, and the service times are exponentially distributed with intensity μ_1 and μ_2 respectively. A call of stream one requires one unit channel and a call of stream two requires two unit channels. Fig. 1 shows the state transition diagram for a system with limited accessibility, where state (i, j) denotes the state of the system (i.e., i and j are the number of unit channels occupied by stream one and two respectively), and $1 - b_{x,d} = \prod_{j=x}^{x+d-1}(1 - b_j)$ is the state dependent acceptance probability derived above. From the figure, we can see that the diagram is reversible as the flow clockwise is equal to the flow counter-clockwise (Kolmogorov's criteria), but there is no product form. Due to reversibility, we can apply the local balance equations to calculate the relative state probabilities, all expressed with reference to state $(0,0)$, then normalize the relative state probabilities to obtain the absolute state probabilities and the relevant performance measures.

Delbrouck [7] developed a general algorithm for calculating the global state probabilities for multi-rate loss systems with BPP-traffic, which is insensitive to

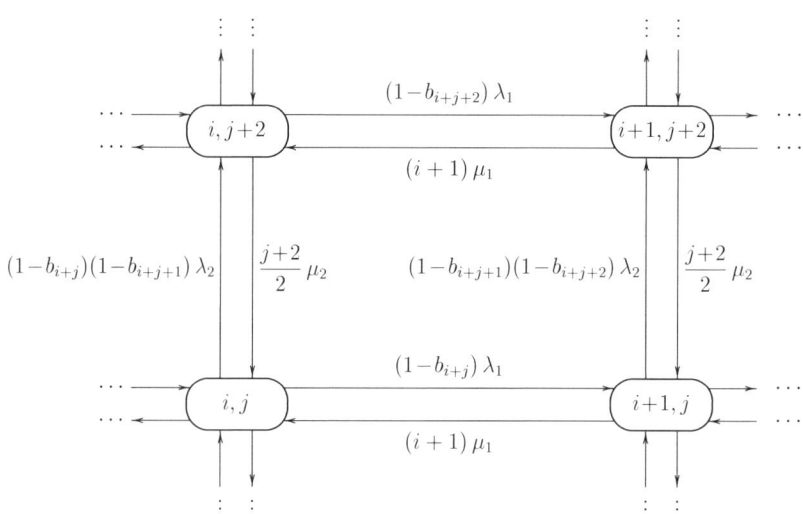

Fig. 1. State-transition diagram with state-dependent blocking probabilities for a multi-rate loss system. The process is reversible as the flow clockwise equals the flow counter-clockwise [5].

service time distribution, i.e., the state probabilities of the system only depend on the holding time distribution through its mean value. Reference [5] extended the Delbrouck's algorithm to allow for evaluating individual performance measures for each service and include state-dependent blocking as shown in Fig. 1. If we consider a system with C unit channels and L traffic streams, the relative global state probabilities $q(x)$ for multi-rate loss systems with state-dependent blocking can be calculated by a generalized recursion formula expressed as follows [5]:

$$q(x) = \begin{cases} 0 & x < 0 \\ 1 & x = 0 \\ \sum_{i=1}^{L} q_i(x) & x = 1, 2, \cdots, C \end{cases} \quad (12)$$

where

$$q_i(x) = \left\{ \frac{d_i}{x} \cdot \frac{A_i}{Z_i} \cdot q(x - d_i) - \frac{x - d_i}{x} \cdot \frac{1 - Z_i}{Z_i} \cdot q_i(x - d_i) \right\} \cdot (1 - b_{x-d_i, d_i}) \quad (13)$$

In the above equations, $q_i(x)$ is the contribution from traffic stream i to the global state $q(x)$. The initialization values of $q_i(x)$ are $\{q_i(x) = 0, x < d_i\}$. The absolute global state probabilities $p(x)$ and $p_i(x)$ can be obtained after normalization.

$$p(x) = \frac{q(x)}{\sum_{j=0}^{C} q(j)} \qquad 0 \le x \le C$$

$$p_i(x) = \frac{q_i(x)}{\sum_{j=0}^{C} q(j)} \qquad 1 \le x \le C \quad (14)$$

A numerically stable algorithm is obtained by normalizing in each step of the iteration [4].

3.3   Performance Measures

Based on the global state probabilities derived above, we are able to get the performance measures of the system in terms of time congestion, call congestion, and traffic congestion.

Time Congestion is by definition equal to the proportion of time the system is blocked for new call attempts. In multi-class OFDM-TDMA systems with AMC scheme, a call attempt of stream i will experience congestion with probability b_{x,d_i} if the system is in state x. Thus the time congestion E_i of stream i is calculated as follows:

$$E_i = \sum_{x=0}^{C} b_{x,d_i} \cdot p(x) \qquad i = 1, 2, \cdots, L \quad (15)$$

Traffic Congestion is by definition equal to the proportion of offered traffic which is blocked. It should be noticed that traffic congestion is the most important performance measure. The carried traffic Y_i of stream i measured in unit channels is given by [5]:

$$Y_i = \sum_{x=1}^{C} x \cdot p_i(x) \qquad i = 1, 2, \cdots, L \tag{16}$$

The offered traffic of stream i measured in unit channels is $d_i \cdot A_i$. Thus the traffic congestion C_i of stream i becomes:

$$C_i = \frac{d_i \cdot A_i - Y_i}{d_i \cdot A_i} \qquad i = 1, 2, \cdots, L \tag{17}$$

Call Congestion is by definition equal to the proportion of call attempts which are blocked. It is said in reference [4] that the call congestion B_i of stream i always can be obtained from the traffic congestion C_i as follows:

$$B_i = \frac{C_i}{Z_i + (1 - Z_i)C_i} \qquad i = 1, 2, \cdots, L \tag{18}$$

where Z_i is the peakedness of stream i.

4   Numerical Results

In this section, we present some numerical results based on the analytical models developed above. We consider the downlink of an OFDM-TDMA system with time division duplex (TDD) operation. The total bandwidth is set to be 5 MHz, which is divided into 5 subchannels with the assumption that the average SNR $\overline{\gamma}_m$ of each subchannel is the same. The duration of a frame is set to be 1 ms so that the channel quality of each connection almost remains constant within a frame, but may vary from frame to frame. The total number of time slots used for downlink data transmission within a frame is set to be 200, each of which contains 4 OFDM symbols. We set the number of transmission modes to 5 with the target PER at 10^{-4}. Three types of service classes with traffic parameters shown in Table 2 are considered. The target outage probability is set to be 2%, which is usually considered to be an acceptable QoS requirement.

The results are shown in Table 3. From the table, we can see that for Poisson traffic ($Z = 1$), the time congestion, call congestion and traffic congestion are identical. This is in accordance with the PASTA property. For Engset ($Z < 1$) and Pascal ($Z > 1$) traffic, there is a difference between the three performance

Table 2. Traffic parameters for the case considered. A is the offered traffic in Erlangs, Z is the peakedness, and d is the bandwidth requirement in unit channels.

Class	Offered traffic	Peakedness	Req. bit rate	Channel bit rate	Channels
i	$d_i \cdot A_i$	Z_i	r_i	r_{unit}	d_i
1	20	1	50 bits/frame	50 bits/frame	1
2	30	0.5	100 bits/frame	50 bits/frame	2
3	30	2	50 bits/frame	50 bits/frame	1

Table 3. Performance measures for the parameters given in Table 2 with state-dependent blocking

Class	Time cong. E	Call cong. B	Traffic cong. C	Carried Traffic Y
1	0.0557	0.0557	0.0557	18.8854
2	0.1132	0.1040	0.0549	28.3544
3	0.0557	0.0604	0.1139	26.5824

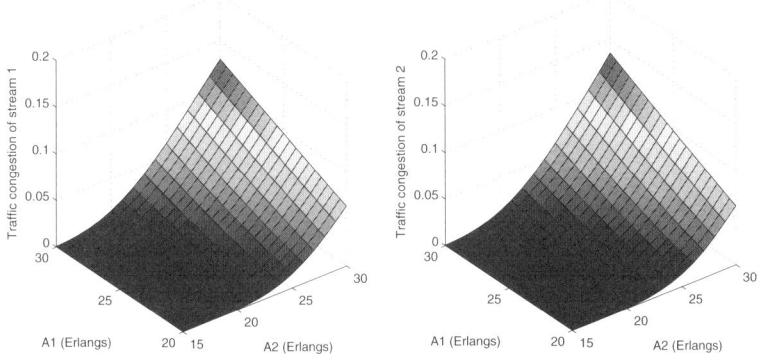

Fig. 2. Traffic congestion versus different traffic loads

measures. From traffic engineering point of view, the traffic congestion is the most important one.

Next, we analyze how different traffic characteristics and various traffic loads will affect the performance metrics. The first two traffic classes in Table 3 are considered, which correspond to voice and data traffic respectively. Fig. 2 shows the traffic congestion versus different traffic loads in Erlangs of the two classes. From the figure, We can see that the increase of the traffic load in class two will increase the traffic congestion in both classes more sharply than class one does. This is because class two requires higher data rate than class one, thus occupies more system resources in terms of time slots.

5   Conclusions

Adaptive modulation and coding (AMC) has been widely used to match transmission parameters to time-varying channel conditions. In this paper, we have developed an analytical model of multi-rate loss system with state-dependent blocking to evaluate the teletraffic performance of multi-class OFDM-TDMA systems with AMC scheme. With state-dependent blocking, the process is still reversible, but the product-form is lost. It has been shown that a d-channel call has the same blocking probability as d consecutive single-channel calls. So the blocking probability of a call attempt depends both on the state of the system and the bandwidth required. Due to reversibility, we have local balance and may calculate the global state probabilities by an effective algorithm. The performance measures in terms of time congestion, call congestion and traffic congestion are derived with numerical examples.

References

1. Qingwen, L., Shengli, Z., Georgious, B.: Queuing With Adaptive Modulation and Coding Over Wireless Links: Cross-Layer Analysis and Design. IEEE Transactions on Wireless Communications 4(3), 1142–1153 (2005)
2. Tarhini, C., Chahed, T.: System capacity in OFDMA-based WiMAX. In: International Conference on Systems and Networks Communication, ICSNC 2006, vol. 4(3), pp. 70–74 (2006)
3. Wang, H., Iversen, V.B.: Erlang Capacity of Multi-class TDMA Systems with Adaptive Modulation and Coding. In: The IEEE International Conference on Communications (ICC 2008), Beijing, China (accepted, 2008)
4. Iversen, V.B.: Reversible Fair Scheduling: The Teletraffic Theory Revisited. In: Mason, L.G., Drwiega, T., Yan, J. (eds.) ITC 2007. LNCS, vol. 4516, pp. 1135–1148. Springer, Heidelberg (2007)
5. Iversen, V.B.: Modelling Restricted Accessibility for Wireless Multi-service Systems. In: Cesana, M., Fratta, L. (eds.) Euro-NGI 2005. LNCS, vol. 3883, pp. 93–102. Springer, Heidelberg (2006)
6. Iversen, V.B.: Teletraffic Engineering Handbook, COM department, Technical University of Denmark, 336 p. (2005)
7. Delbrouck, L.: On the Steady-State Distribution in a Service Facility Carrying Mixtures of Traffic with Different Peakedness Factors and Capacity Requirements. IEEE Transactions on Communications 31(11), 1209–1211 (1983)
8. Sarkar, T.K., Zhong, J., Kyungjung, K., Medouri, A., Salazar-Palma, M.: A survey of various propagation models for mobile communication. IEEE Antennas and Propagation Magazine 45(3), 51–82 (2003)

Reducing Problems in Providing Internet Connectivity for Mobile Ad Hoc Networks*

Quan Le-Trung and Gabriele Kotsis

Department of Telecooperation, Johannes Kepler University
Altenberger Str. 69, A4040 Linz, Austria
`{quanle,gk}@tk.uni-linz.ac.at`
`http://www.tk.uni-linz.ac.at`

Abstract. This paper first introduces the required functions on providing Internet connectivity for MANETs and its mobility management. Then two discovered problems are classified and discussed. They include *inconsistent context* and *cascading effect*. The former appears due to the mobility of ad hoc nodes among multi-homed Internet domains, the ad hoc routing forwarding strategies, and the requirement of backward/ forward traffic of 2-way connection, e.g. TCP, which must be transmitted through the registered Internet gateways. The latter is due to the determination of MANET node location using re-active ad-hoc routing protocols. Finally, solutions are proposed, which either reduce partly or eliminate completely discovered problems.

Keywords: Inconsistent context, cascading effect, Internet connectivity, Internet gateway, MANET, ad-hoc routing, multi-homed domain.

1 Introduction

Attempts are in progress to connect mobile ad hoc networks (MANETs) to the Internet infrastructure to fill in the coverage gaps in the areas where the first-hop coverage, e.g., WLANs or cellular networks, is not available. In the very near future, mobile nodes will roam across multiple heterogeneous platforms while continuously maintaining connection connectivity. A mobile node may connect to a WLAN, and then move into an area where the coverage from the WLAN does not exist. There, it may reconfigure itself into ad hoc mode and connect to a MANETs. Essential to such seamless mobility is efficient mobility management and handoff support, as well as forwarding traffic strategies out/in the Internet. For the MANET integration with IP networks, the MANET should adapt to the network functionality within IP networks, and it is usually considered as the complementary to the IP network, where the Internet connectivity is extended into a MANET, making a MANET a part of the Internet [1]-[2].

The current trend in *Internet mobility management* is to use mobile IP (MIP) [3] for the *macro-mobility*, i.e., the mobility of Internet hosts among Internet domains

---

* This work is supported by EuroFGI project in the Department of Telecooperation, Johannes Kepler University Linz.

L. Cerdà-Alabern (Ed.): Wireless and Mobility 2008, LNCS 5122, pp. 113–127, 2008.

belong to different Internet service providers (ISPs). However, .MIP entails high signaling overhead if there is a long distance between the foreign subnet that a mobile node moves in and its home network, as well as the ping-pong effect of mobility. To reduce this overhead, solutions can be generally classified into two approaches: *micro-mobility* [6] and *fast handover* [7]-[8]. The former uses the address of an Internet gateway (IGW) that is common to a potentially large numbers of network access points (APs) or base stations. When a mobile host moves from one AP to another, which is reachable through the same IGW, there is no need to inform its home agent (HA), and thus reducing the signaling overhead. The latter reduces delay and packet loss during the handoff by using the coupling and synchronization between IP and radio layer to predict the new subnet the mobile host will move into.

Since the connection between a MANET node and an IGW is multi-hop, there is no direct wireless link from this MANET node to the IGW, but they are connected via other immediate nodes. Thus, a MANET node cannot initiate handoffs that are based on the link quality to the AP, i.e., fast handover. Therefore, the trend in *MANET mobility management* is to use MIP [3] for the macro-mobility and MANET routing protocols [9] (proactive vs. reactive vs. hybrid) for the micro-mobility [1].

In this paper, we consider two different problems, called *inconsistent context*, and *cascading effect*. The *former* appears due to the mobility of MANET nodes among multi-homed Internet domains, the ad-hoc routing forwarding strategies, and the requirement of backward/ forward traffic of 2-way connections, e.g., TCP, which must be transmitted through the registered IGWs. Otherwise, the connection is terminated. This type of traffic is generated by either a MANET node or an Internet host, forwarded to the other via multiple IGWs and network address translation (NAT) devices, which locate between the MANET and the Internet. The appearance of multiple IGWs, NAT devices, together with their ingress filtering policies, introduces different approaches for associating[1] MANET nodes in one domain (subnet) to different IGWs, and (re)-registering to their HAs. This creates the misunderstanding between which IGW a MANET node associates or (re)-registers, and which IGW forwards actually traffic for the MANET node, called *inconsistent context*. This situation can happen on any multi-hop chains where different MANET nodes on the chain update their own views (contexts) for forwarding traffic in/out the Internet during the operation of corresponding MANET routing protocol (proactive vs. reactive vs. hybrid) [9] and IGW discovery method (proactive vs. reactive vs. hybrid) [1], [11]-[12]. The *latter* is due to the determination of MANET node location using re-active ad-hoc routing protocols.

This paper is organized as follows. Part 2 continues with the required functions of providing Internet connectivity for MANETs and the mobility management, together with related works. Part 3 shows different realistic scenarios, in which the inconsistent context and cascading problems appear. Solutions are then presented in part 4, which either reduce partly or eliminate completely discovered problems. Part 5 ends this paper with conclusions.

---

[1] In scope of this paper, the association means a MANET node chooses an IGW for its outbound traffic to the Internet, but this IGW can be different from the IGW it registers with its home agent (for inbound traffic from the Internet), e.g., for updating to the shorter route to another IGW. As a result, this creates the *inconsistent context* problems as described in the subsequent parts.

2 Required Functions on Providing Internet Connectivity for MANETs and Mobility Management

In this section, the needed functions of providing Internet accesses for MANET nodes and mobility management are described. They include: *MANET node location determination*, *IGW discovery*, *IGW selection*, *IGW forwarding strategy*, *address auto-configuration*, and *handoff-style*. Related works are also discussed following the descriptions of above functions.

MANET node location determination: A destination node can be determined to be either within the same MANET domain of the source MANET node or an Internet host, by one of the following methods [1], [15]:

- *Network prefix*: Each MANET node must be assigned a global unicast IP address, both home address (HAddr) and care-of address (CoA), i.e., MANET node address is topologically correct [15].
- *Routing table* in MANET proactive ad-hoc routing protocols [9]: if an entry for the destination is in the routing table of the source MANET node, the destination is either in the same MANET domain (subnet), or an Internet host via the IGW. Otherwise, the destination is unreachable.
- *Flooding* route request (RREQ) and waiting for route reply (RREP) in MANET reactive ad-hoc routing protocols [9]: if a host route is returned, the destination is in the same MANET domain (subnet). If a default route is returned, the destination is either an Internet host via the IGW, or unreachable.
- *Internet gateway*: In responding to a RREQ, sending a proxy RREP to signal that it can route to requested destination, i.e., analogous to functionality of a proxy ARP, but over the multi-hops. To do this, an IGW must determine that the destination is not in the same MANET by keeping a list of currently known active nodes (called *visitor list*), or pinging destination on the IGW network interface attaching to the Internet, or flooding the whole MANET with a new RREQ [1], [15].

Internet gateway discovery: To access Internet, each MANET node needs to determine an IGW to forward its traffic to the Internet and vice verse. Different mechanisms can be used to discover the IGW, which can be classified into three sub-classes: *proactive*, *reactive*, or *hybrid* [11]. In the *proactive* approach, each IGW broadcast periodically the advertisement[2], while a MANET node sends a solicitation and waits for a reply from the IGW in the *reactive* approach. The former takes much overhead in MANET, while the latter entails the longer delay. The *hybrid* approach compromises with the balance, in which each IGW broadcast periodically the advertisement within the radius of n-hop. MANET nodes, which are located further n-hop from the IGW, use the reactive approach for IGW discovery [12].

Internet gateway selection: If a MANET node discovers multiple IGWs for accessing to the Internet, a metric is needed to select the right one. Different metrics can be used:

---

[2] It is agent advertisement in MIPv4, or router advertisement in MIPv6.

- *Shortest hop-count* to the nearest IGW [15].
- *Load-balancing*: For intra-MANET traffic, choosing different immediate relays node to the destination MANET nodes within the same MANET domain, while for inter-MANET/Internet traffic, choosing different IGWs for forwarding traffic from/to the MANET to/from the Internet [13].
- *Service class*: Depend on the service classes provided and supported by each IGW.
- *Euclidean distance*: Distance (space or hop-count) between the MANET node and the IGW [14].
- *Hybrid*: A combination of some of above metrics [14].

Internet gateway forwarding strategies: This function takes the responsibility to forward traffic in/out the Internet, or within the MANET. Typically, it can be classified into *inter-MANET* and *intra-MANET forwarding strategies*. The *former* uses different approaches as follows:

- *Default routes*: Represent the default next-hop to send packets to that do not match any other explicit entry in a MANET node's routing table. Usually, the default route is used to send packets to the IGW, where packets are forwarded to the destination host in the Internet [15]-[16].
- *Tunneling* (or encapsulation): Usually, IP-in-IP encapsulation technique is used for traffic in/out the Internet. Outer IP header is for the connection between the source MANET node and the IGW, i.e., tunneling, while inner IP header is for the connection between the source MANET node and the destination [5].
- *Half-tunneling*: Traffic to the Internet from the MANET domain uses tunneling, while traffic from the Internet to the MANET domain uses the ad-hoc forwarding [5].
- *Source routing*: A list of all immediate nodes between the source MANET node and the IGW are added into the IP header. At the IGW, the source routing header is removed and the packet is continuing forwarded as a normal packet [17].
- *Spanning tree* rooted at the IGW: Taking the benefit of proactive IGW discovery. A tree rooted at the IGW is built and maintained using the agent advertisements broadcasted periodically by the corresponding IGW [18].

The *latter* is totally based on the operation of ad-hoc routing protocols, which can be classified as *proactive*, or *reactive*, or *hybrid* [9]. In *proactive* approach, each node continuously maintains up-to-date routing information to reach every other node in the network. Routing table updates are periodically transmitted throughout the network in order to maintain table consistency. Thus, the route is quickly established without any delay. However, for a highly dynamic network topology, the proactive schemes require a significant amount of resources to keep routing information up-to-date and reliable. In *reactive* approach, a node initiates a route discovery throughout the network, only when it wants to send packets to its destination. Thus, nodes only maintain the routes to only active destinations. A route search is needed for every new destination. Therefore, the communication overhead is reduced at expense of delay due to route research. Finally, in *hybrid* approach, each node maintains both topology information within its zone via the proactive approach, and the information regarding neighbor zones via the reactive approach.

Address auto-configuration: In order to enable a MANET to support IP services and the internetworking with the Internet, MANET address space based on IPv4/IPv6 is required. Moreover, the MANET addressing schemes must be auto-configured and distributed to support for the self-organized and dynamic characteristics of MANETs. Numerous addressing schemes for MANETs based on IP address auto-configuration have been proposed in the literature. They can be classified into two approaches [20]: *conflict-detection allocation* and *conflict-free allocation*.

In the *former*, mechanisms are based on picking an IP address from a pool of available addresses, configuring it as tentative address and asking the rest of the nodes of the network, checking the address uniqueness and requesting for approval from all the nodes of the network. In case of conflict, e.g., the address has been already configured by another node, the node should pick a new address and repeat the procedure (sort-of "trial and error" method). This process is called duplicate address detection (DAD). In the *latter*, mechanisms assume that the addresses that are delegated are not being used by any node in the network. This can be achieved, for by ensuring that the nodes that participate in the delegation have disjointed address pools. In this way, there is no need of performing the DAD procedure.

Although an IP-based address auto-configuration scheme is preferred in self-organizing the MANETs for their fast deployments, only stateless mechanism is suitable for MANETs [19]. This is because the stateful mechanism requires a centralized server to maintain a common address pool, while the stateless mechanism allows the node to construct its own address and is suitable for MANETs. However, in the conflict-detection allocation, a DAD mechanism is required to assure the uniqueness of the address, especially to support for MANET merging and partitioning.

Finally, the address allocation space is important. It must be large enough to cover the large-scale MANETs and reduce the probability of address conflicts. The following IPv4 and IPv6 addressing spaces have been proposed for MANETs [15]: *169.254.0.0/16* for IPv4, and *FEC0:0:0:FFFF::/64 (MANET_PREFIX)* for IPv6.

Handoff-style: A node performs a handoff if it changes its IGW while communicating with a correspondent node (CN) in the Internet.

In conventional mobile networks, e.g., WLAN, the quality of the wireless link between mobile node and APs determines when to handoff from one AP to another. The performance of these types of handoffs depends on the mobility management protocol in the access network. In MANETs, the situation is more complicated. In general, some nodes do not have a direct wireless link to an AP, but they are connected via other immediate nodes. Thus, they cannot initiate handoffs that are based on the link quality to the AP. Rather, the complete multi-hop path to the AP, which serves the current IGW, must be taken into consideration. A handoff can occur if an ad-hoc node itself or any of the intermediate relay ad-hoc nodes moves and breaks the active path. In general, if the path between an ad-hoc node and the IGW breaks and there is no other path to the same IGW, the ad-hoc node has to perform the IGW discovery to establish a new path to another IGW [21].

The IGW discovery scheme and the ad hoc routing protocol both have huge influence on the multi-hop handoff performance. Multi-hop handoff schemes can be classified into *forced handoff* and *route optimization-based handoff*.

The *former* occurs whenever the path between the source/destination mobile node and the IGW is disrupted during data transmission due to some reasons, e.g., the movement of the MANET node. Therefore, a new path to the Internet has to be set up. The following IGW discovery process may result in the detection of a new IGW, which will consequently result in a handoff. The *latter* is a handoff as a result from route optimization. If the source/destination MANET node detects that a shorter path to the Internet becomes available while communicating with a corresponding node, the active path will be optimized. In case the shorter path is via a different IGW, a *route optimization-based handoff* occurs.

3   Scenarios of Inconsistent Context and Cascading Problems

This part continues describing problems on providing Internet connectivity for MANETs and the mobility management in realistic scenarios shown in Fig. 1, of which required functions have already been presented in part 2. In this section, we assume that MANET mobility management uses MIPv4 for the *macro-mobility* and ad hoc routing protocol for the *micro-mobility*. The data traffic sent from a MANET node to an Internet host is forwarded through IGWs using either the *default route* or the *tunneling techniques* [1], [4]-[5]. There are also multiple IGWs, NAT devices, together with their ingress filtering policies between the MANET domain and the Internet. Fig. 1 shows different scenarios on IGW forwarding strategies, which are used to illustrate for different discovered problems next.

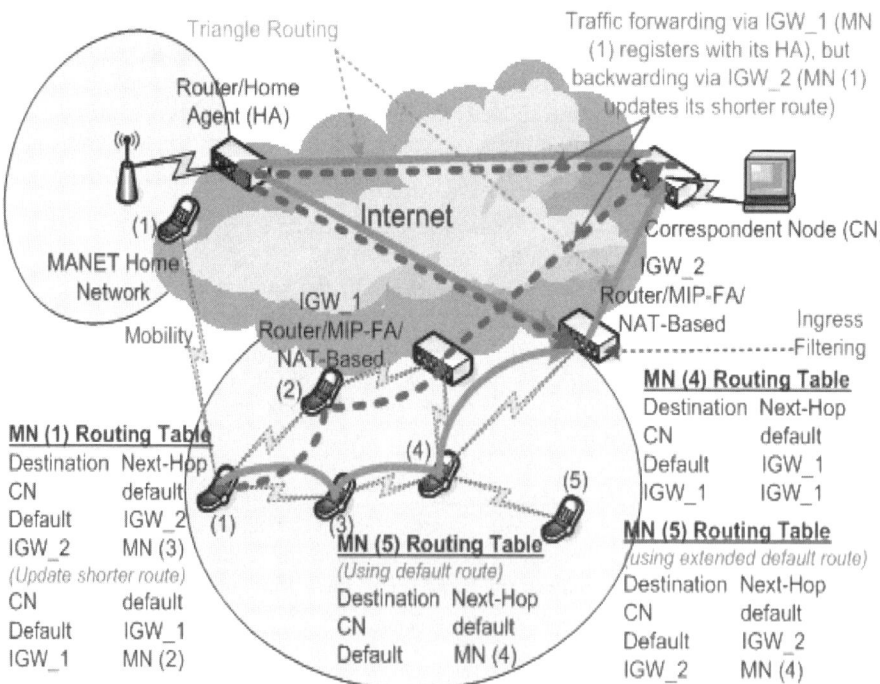

Fig. 1. Different problems on Internet gateway forwarding strategies

3.1   Inconsistent Context due to Default Route Forwarding (Type I)

In this forwarding mode, a MANET node (MN) sends packets to the Internet, i.e., communicate with the correspondent node (CN), using the *default route*. Whenever it associates with an IGW, it sets its *default route* pointing to this IGW. The *default route* is used to forward data packets to the unknown destination. In the scope of MANET, it usually means that the destination is not in the same MANET domain and can be located in the Internet, thus the data packet is forwarded to the IGW, where it is dropped if the destination is unreachable, or continued forwarding to the Internet if the destination is reachable via IGW.

MN (5)'s routing table using default route in Fig.1 shows an example, where MN (4) is used as the next-hop to the associated IGW_1.

The problem with this *default route* setting is that MN (5) does not know its current associated IGW. Thus, the *inconsistent context* on sending packets from MN (5) to the Internet, e.g., via IGW_2, and receiving packets from CN to MN (5), e.g., via IGW_1, can terminate the 2-way connection like TCP. The next section shows an advanced setting to reduce this effect.

3.2   Inconsistent Context due to Default Route Forwarding (Type II)

In this forwarding mode, a MANET node adds an additional entry into its routing table for the default route, indicating the current IGW that this MANET node associates. Fig. 1 shows MN (5)'s routing table *using extended default route*.

With this advanced setting, the ambiguous IGW association of MANET node is solved. However, an additional cost on storage and access time for the additional entry is introduced. Moreover, the *inconsistent context* problems are still existed in other scenarios, which are presented in the next section.

3.3   Inconsistent Context due to Default Route Forwarding (Type III)

Different scenarios indicating the *inconsistent context* are analyzed in this section, which can be dependent on the operation of MANET routing protocols [9] and IGW discovery methods (proactive vs. reactive vs. hybrid) [1], [11]-[12].

Scenario I:

A MANET node updates a shorter route to another IGW without re-registering the new IGW with its home agent (HA), and/or its foreign agent (FA), as well as the NAT device located in the visiting MANET domain.

Fig. 1 shows an example, where MN (1) moves from its home network to the new MANET domain, registering to its home agent (HA) via IGW_2 through MN (3) as the next-hop to the IGW_2. The distance from MN (1) to IGW_2 is 3-hop. Later, MN (1) finds a shorter route to the Internet via IGW_1. MN (1) chooses MN (2) as the next-hop to the IGW_1 and the hop-count is 2. In this scenario, traffic from MN (1) to the CN is forwarded via IGW_1. However, traffic from the CN to MN (1) is still forwarded via IGW_2 since it is registered by MN (1) to MN (1)'s HA. This creates the *inconsistent context*.

Scenario II:

A MANET node associated with an IGW, e.g., IGW_1, forwards agent advertisement packet for another IGW, e.g., IGW_2. As a result, its downstream nodes can associate to the IGW_2 using it as the next-hop. However, the traffic is actually forwarded through IGW_1.

As an example, in Fig. 1, suppose that MN (4) is currently associates with IGW_1 and MN (5) is not associated with any IGWs. In the proactive IGW discovery [1], the IGW will broadcast periodically its agent advertisement packets. When MN (4) receives IGW_2 agent advertisement packets, MN (5) will set MN (4) as the next-hop to IGW_2 in its default route, if MN (4) continues forwarding IGW_2's agent advertisement packets. However, whenever the traffic to CN is generated by MN (5), and then forwarded to IGW_2 via MN (4), this traffic is actually forwarded via IGW_1 by MN (4). This creates the *inconsistent context*.

Note that the decision of MN (4) whether or not to continue forwarding other IGWs packets, e.g., agent advertisement, is dependent on the corresponding operation and implementation of MANET routing protocol [9] and IGW discovery method [1].

Scenario III:

A MANET node looses its association to the current IGW, e.g., a detection of broken link, and re-associates to another IGW. As a result, traffic to Internet from its downstream nodes choosing it as the next-hop to the current IGW will be forwarded via another IGW instead.

As an example, in Fig. 1, suppose MN (3) chooses MN (4) as its next-hop to the IGW_1 and MN (4) is currently associated to the IGW_1. Later, if the link between MN (4) and IGW_1 is broken, MN (4) re-associates to IGW_2. Thus, traffic to the Internet from MN (3) will be forwarded via IGW_2, though MN (3) thinks it is forwarded via IGW_1. This creates the *inconsistent context*.

3.4   Inconsistent Context due to MIPv4-FA Triangle Routing Problem

The use of MIPv4 can easily lead to *triangle routing problem*, see Fig. 1 in green arrows, i.e., traffic from a MANET node to the correspondent node (CN) is sent directly, while return traffic is sent to the MANET node's home agent (HA), then is tunneled to MANET node's foreign agent (FA) and delivered to the MANET node.

On providing Internet connectivity for MANET nodes, this also means that traffic to the Internet from a MANET node can be forwarded to one IGW, e.g., IGW_1 in Fig. 1, for a shorter route, while return traffic is forwarded through registered IGW, e.g., IGW_2 in Fig. 1. This creates the *inconsistent context*. In the case IGW_2 is behind a stateful firewall, i.e., the first outbound to Internet packet sets the soft state in the firewall for the return packets to enter the MANET domain, the connection will be terminated. The pink arrows in Fig. 1 show an example.

Note that *triangle routing problem* can be considered as a consequence of Scenario I in Part 3.3, so solution for this problem can be referred in Scenario I, Part 4.1.

3.5 Inconsistent Context due to MIPv4-FA Ingress Filtering

Ingress filtering means that a router/firewall will not accept on its ingress interface packets with a source IP address that is not topologically correct for that interface. The motivation is to prevent IP address spoofing. If the *ingress filtering* is integrated into IGWs, the traffic to the Internet from a MANET node ends up at another IGW than the IGW it is registered with, that IGW can drop its packet.

As an example, in Fig. 1, MN (1) moves to a new MANET domain, getting a new topologically correct care-of address (CoA) and registering it with its HA via IGW_2. If MANET node updates later its shorter route through IGW_1, e.g., using another IP subnet, its outbound packets can be dropped at IGW_1 due to *ingress filtering*.

Note that *ingress filtering problem* can be also considered as a consequence of Scenario I, Part 3.3, so solution for this problem can be referred in Scenario I, Part 4.1.

3.6 Inconsistent Context due to MIPv4-FA Traversing NAT

If the IGWs connecting MANET domains with the Internet are either behind the NAT devices, or integrating the NAT function, and private IP subnets are used with MANET domains, a MANET node communicates with an Internet host through the private IP address of its associated IGW, which will be mapped into another public IP address. Thus, with *a NAT-based IGW*, the connection of a MANET node is bound to the NAT that the connection passes through. However, whenever the MANET node updates its new NAT-based IGW, the return traffic can be dropped at this one due to the stateful router/firewall. This creates the *inconsistent context*. Solution for this problem is shown in Part 4.3.

3.7 Cascading Effect in MANET Node Location Determination

On reactive ad-hoc routing protocol, whenever a route to a destination is needed, source MANET node needs to perform a route discovery. Usually, it broadcasts a route request (RREQ) packet and waiting for a route reply (RREP), if the destination is located in the same MANET. On providing Internet connectivity for MANET nodes using the default routes, no RREP will be sent back to the source MANET node if the destination is an Internet host. Note that for some mechanisms, it is allowed for an IGW to send a RREP to the source MANET node to indicate that the destination is an Internet host and can be reached through this IGW, called *proxy RREP*.

If the destination is an Internet host, the source MANET node will send packets to the destination using its default route setting in its routing table. However, when its upstream MANET node receives its packets, its upstream MANET node again performs another route discovery for the destination. This route discovery is repeated at each upstream MANET node on the chain from the source MANET node to the corresponding IGW, creating a problem called *cascading effect*.

The advantage of *cascading effect* on MANET reactive routing protocols is that it always found the shortest host route to the destination if the destination is located within the same MANET domain. However, this mechanism is redundant and entails much overhead if the destination is an Internet host.

Therefore, *cascading effect* problem should be removed on providing Internet connectivity for MANET nodes. One solution is to insert directly in the routing table of each MANET node an entry for each destination Internet host. When a relay MANET node receives data packets from its downstream nodes, it continues forwarding to the destination Internet host without any route discovery, if an entry for that destination Internet host is available in its routing table.

4 Solutions

Since the *inconsistent context problems* due to the MIPv4-FA triangle routing, ingress filtering are consequences of Scenario I, Part 3.3, and the solution for *cascading effect* problem is already presented, this part continues with only solutions for the *inconsistent context* due to *default route forwarding* in Part 4.1 and Part 4.2, as well as solution for the *inconsistent context* due to *MIPv4-FA traversing NAT* in Part 4.3.

4.1   Reducing Inconsistent Context of Using Default Route (Type III)

Scenario I:

A MANET node is not allowed to update its shorter route to another IGW, unless its current transmissions on any 2-way connections to the Internet hosts are finished, and it has already re-registered this new IGW with its home agent. This re-registration can be prepared in advance, e.g., during the data transmissions on the current connections via the old IGW.

Clearly, this rule removes completely the *inconsistent context* in Scenario I since the inbound/outbound traffic to/from Internet is always forwarded via the same IGW. However, its disadvantage is that intra-MANET route from the source MANET node to the registered IGW is not an optimal one, e.g. the shortest route.

This rule also reduces the *inconsistent context* in Scenario II and Scenario III. The reason is that, the less changing on the IGW re-association of MANET nodes, the less *inconsistent context* problems appear on their downstream MANET nodes.

Scenario II:

On proactive IGW discovery, a MANET node that does not register to any IGW is allowed to re-broadcast the received agent advertisement if it decides to register with this IGW. Otherwise, the re-broadcasting of agent advertisement is prohibited.

On re-active IGW discovery, a MANET node is allowed to generate/forward an agent advertisement in one of the following three cases:

- *It does not register to any IGW, registering it to the IGW of which the agent advertisement it receives, then forwarding the agent advertisement to the source MANET node.*
- *It has already registered to an IGW, receiving the agent advertisement generating by the same IGW, forwarding this agent advertisement to the source MANET node.*
- *It has already registered to an IGW, receiving the agent solicitation from the source MANET node, generating itself an agent advertisement to the source MANET node.*

On hybrid IGW discovery, the above rules are applied whenever an agent advertisement or an agent solicitation is received.

Clearly, the above rules ensure that all the MANET nodes on the chain from any source MANET node to their registered IGWs uses the same IGW for inbound/outbound traffic to/from the Internet. However, on the proactive IGW discovery, the applied rule creates non-overlapped MANET domains, each domain associates to only one IGW. Thus, it does not take the advantage of using multiple IGWs for load balancing and fault-tolerance. Moreover, there can be the appearance of orphan MANET nodes due to the collision or high mobility.

On reactive or hybrid IGW discovery, the generation of agent advertisement of an IGW or an immediate MANET node, called *gratuitous agent advertisement* or *proxy route reply* (*RREP*), if this agent advertisement is piggybacked on *RREP* of MANET reactive routing protocol, to the source MANET introduces non-optimal route. This happens whenever the destination is located in the same MANET domain of the source MANET node, and the source MANET node receives a *proxy RREP* before the normal *RREP* sent by the destination, e.g., due to the collision or longer hop-count. This problem can be further reduced by the use of the destination sequence number, in which this field on the *proxy RREP* packet is always set to a small value, e.g., *"0"*. Thus, when the source MANET node receives the *RREP* sent by the destination later, it will updates the host route to the destination instead of the default route via the IGW, since the destination sequence number sent by the destination is greater than that of *proxy RREP* sent by the IGW.

Scenario III:

There should be a mechanism for the MANET node detected the broken link, sending this information to its downstream MANET nodes so that these MANET nodes can re-register their new IGWs with their home agents. This mechanism is usually MANET routing protocol dependent.

As an example, in ad-hoc on-demand distance-vector (AODV) routing [22], a MANET node keeps a list of other neighbor MANET nodes using it as the next-hop to a set of destinations, called precursor list. Whenever this MANET node detects a broken link to any destination, it searches its precursor list for that destination and sends a route error (RERR) to all the nodes on this list. This process is propagated to the MANET nodes in its precursor list.

In the scope of this scenario, a MANET node detecting a broken link to its registered IGW will integrate this information to the *RERR* and send it to all nodes in its precursor list. However, to reduce also the *inconsistent context* in the **Scenario II**, the detected MANET node also sends the *RERR* to all the precursor lists associated to all IGWs as the destinations. This is because some MANET nodes can be associated with an IGW, but their traffic can be forwarded to another IGW instead.

4.2   Removing Inconsistent Context Using Tunneling

Clearly, on IGW forwarding strategies using *default route* between a MANET domain and the Internet connecting through multiple IGWs, the *inconsistent context* is always a problem. This is because the traffic from the MANET to the Internet and the returned traffic can be forwarded through different IGWs, taking to the termination of 2-way connections.

With the described solutions on the previous section, the problems on *default route* are reduced. However, depending on the implementation of MANET routing protocols [9], the IGW discovery (whether they are implemented independently or dependently) [11], there can be the appearance of another scenarios of *inconsistent context*. Thus, it is the purpose of this part to introduce another approach to remove completely the effect of *inconsistent context*.

The idea is that a MANET node is always sent its outbound traffic to the Internet via its registered IGW, irrespective its immediate MANET nodes updating the shorter routes to other IGWs. This is achieved via the use of *tunneling*, e.g., *IP-in-IP tunneling* [4]-[5]. In this scheme, the source MANET node encapsulates the original IP packet to another IP packet, in which its registered IGW is the destination IP address in the outer IP header. The encapsulated IP packet is then forwarded to its registered IGW using the MANET routing protocol. When the encapsulated IP packet arrives to the IGW, it is decapsulated and the inner original IP packet is forwarded to the destination Internet host by the IGW.

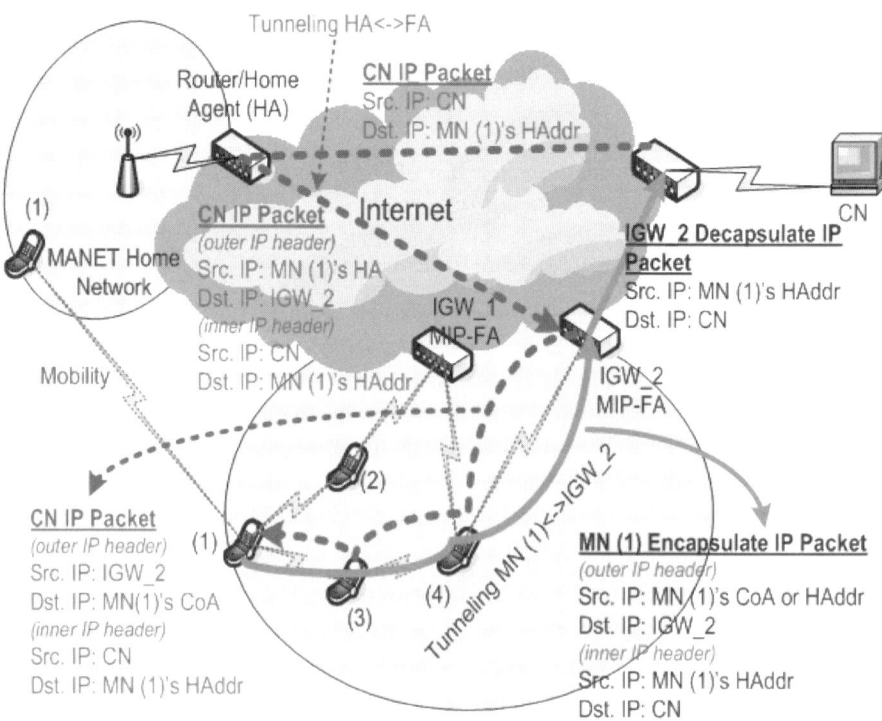

Fig. 2. Packet header passing MIP-FA IGW using tunnelling

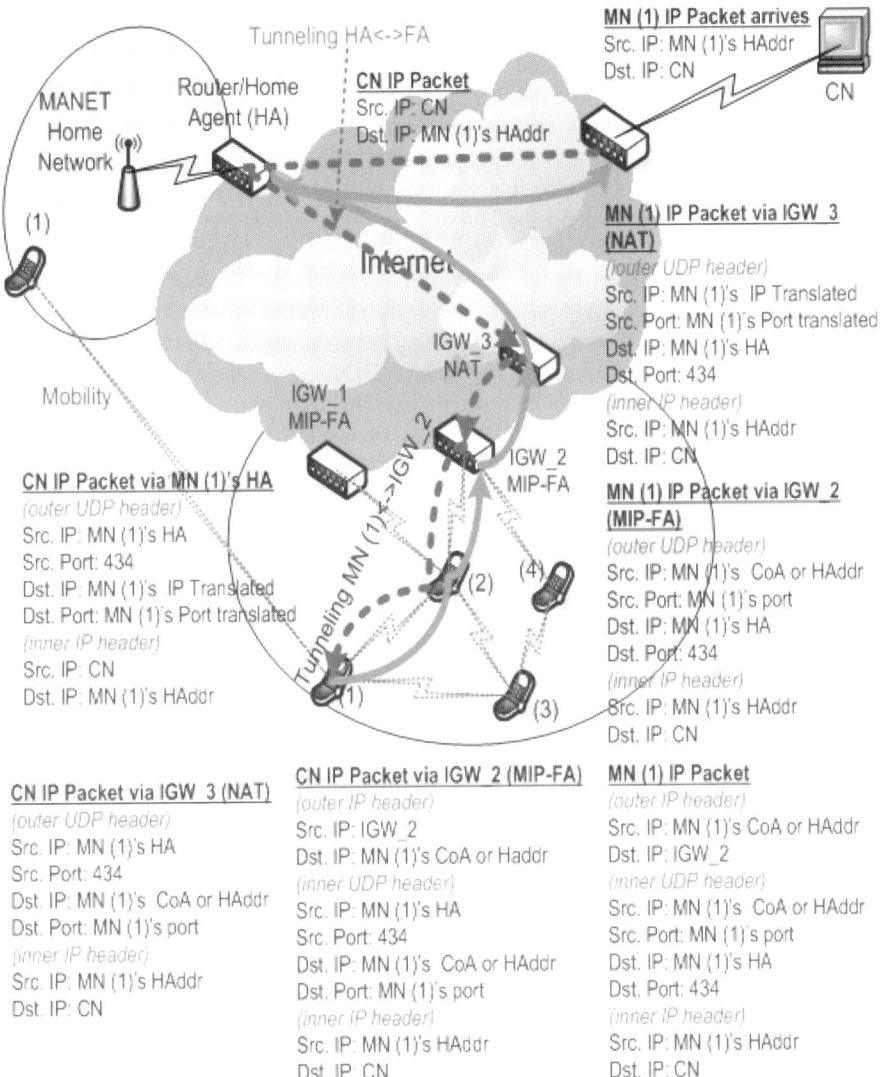

Fig. 3. Packet header passing both MIP-FA and NAT-based IGWs using tunneling

Fig. 2 shows an example, where an IP packet to the Internet from the source MANET node MN (1) is first tunneled (encapsulated) to its registered IGW_2, then it is decapsulated and the original IP packet is forwarded to the CN. In the reversed direction (using tunneling through MN (1)'s HA in this scenario), CN send its packet to the MN (1) home address (HAddr) via MN (1)'s HA, where the packet is encapsulated and forwarded to MN (1)'s FA (IGW_2). At the IGW_2, this packet is decapsulated, then encapsulated and forwarded to the MN (1).

With the tunneling, the *inconsistent context* problems are completely removed. However, an additional cost of adding an outer IP header is introduced.

4.3 Solutions on Passing the NAT Device

Mobile IP with NAT traversal can be used to pass the *NAT-based IGW*. The NAT traversal is based on MIP UDP tunneling mechanism [10].

In MIP UDP tunneling, the mobile node may use an extension in its registration request to indicate that it is able to use MIP UDP tunneling instead of standard MIP tunneling if the home agent sees that the registration request seems to have passed through a NAT. The home agent may then send a registration reply with an extension indicating acceptance (or denial). After assent from the home agent, MIP UDP tunneling will be available for use for both forward and reverse tunneling. UDP tunneled packets sent by the MANET node use the same ports as the registration request message. In particular, the source port may vary between new registrations, but remains the same for all tunneled data and re-registrations. The destination port is always 434. UDP tunneled packets sent by the home agent uses the same ports, but in reverse.

To reduce the *inconsistent context* problems, this solution requires an additional IP-in-IP tunneling from the source MANET node to its registered IGW, to ensure that the source MANET IP address and source MANET UDP port number are translated consistently by the same *NAT-based IGW*.

Fig. 3 shows an example, where the *tunneling* is established between the MN (1) and its registered IGW_2, and between the MN (1)'s MIP-FA (mapping to publish IP address of IGW_3) and MN (1)'s HA. MIP UDP tunneling is used to transmit packet through the NAT-based IGW_3, while the tunneling between MN (1) and IGW_2 is used to remove the *inconsistent context* problems.

Note that in Fig. 3, MN (1) IP Packet is used to indicate the direction of sending data packets from source MANET node to the destination CN, while CN IP Packet is used for indicating the packet transmissions in the reversed direction, respectively.

5 Conclusions

This paper describes different functions which are needed for providing Internet connectivity of MANET nodes and their mobility management. Then two classes of problems, called *inconsistent context* and *cascading effect*, are discussed. The former is mainly due to the ad hoc forwarding strategies through multiple Internet gateways, NAT devices, together with their ingress filtering, using either default route or tunneling techniques. The latter is due to the operation of reactive ad-hoc routing protocols. Corresponding solutions have been also proposed, either reducing partly or removing completely the discovered problems. The solutions can be applied for either the proactive or the reactive MANET routing protocols. They can be also integrated into the MANET routing protocol to reduce overhead, increasing performance, or implemented as a separate protocol.

References

1. Abduljalil, F.M., Bodhe, S.K.: A Survey of Integrating IP Mobility Protocols and Mobile Ad Hoc Networks. IEEE Commu. Surveys and Tutorials 9(1), 14–30 (2007)
2. Le, D., Fu, X., Hogrefe, D.: A Review of Mobility Support Paradigms for the Internet. IEEE Communications Surveys and Tutorials 8(1), 38–51 (2006)

3. IETFs Mobility for IPv4 Charter,
   http://www.ietf.org/html.charters/mip4-charter.html
4. Engelstad, P.E., Tonnesen, A., Hafslund, A., Egeland, G.: Internet Connectivity for Multi-Homed Proactive Ad Hoc Networks. In: IEEE ICC 2004, pp. 4050–4056 (2004)
5. Jönsson, U., Alriksson, F., Larsson, T., Johansson, P., Maguire, G.Q.: MIPMANET – Mobile IP for Mobile Ad Hoc Networks. In: MobiHoc 2000, Boston, Massachusetts, pp. 75–85 (2000)
6. Campbell, A.T., Castellanos, J.-G.: IP Micro-Mobility Protocols. ACM SigMobile MC2R 4(4), 45–53 (2000)
7. Koodli, R., Perkins, C.: Mobile IPv4 Fast Handovers. Internet draft, draft-ietf-mip4-fmipv4-07.txt (2007)
8. Koodli, R.: Fast Handovers for Mobile IPv6. In Internet rfc4068.txt (2005)
9. IETF MANET WG Charter,
   http://www.ietf.org/html.charters/manet-charter.html
10. Levkowetz, H., Vaarala, S.: Mobile IP NAT/NAPT Traversal using UDP Tunneling. In Internet Draft draft-ietf-mobileip-nat-traversal-07.txt (2007)
11. Jin, X., Christian, B.: Wireless Multihop Internet Access: Gateway Discovery, Routing, and Addressing. In: Proceedings of 3GWireless 2002, San Francisco, CA, USA (2002)
12. Ruiz, P.M., Gomez-Skarmeta, A.F.: Adaptive Gateway Discovery Mechanisms to Enhance Internet Connectivity for Mobile Ad Hoc Networks. Ad Hoc & Sensor Wireless Networks 1, 159–177 (2005)
13. Hsu, Y.Y., Tseng, Y.C., Tseng, C.C., Huang, C.F., Fan, J.H., Wu, H.L.: Design and Implementation of Two-Tier Mobile Ad Hoc Networks with Seamless Roaming and Load-Balancing Routing Capability. In: IEEE QSHINE 2004, pp. 52–58 (2004)
14. Ammari, H., Rewini, H.E.: Using Hybrid Selection Schemes to Support QoS When Providing Multihop Wireless Internet Access to Mobile Ad Hoc Networks. In: QSHINE 2004, pp. 148–155 (2004)
15. Perkins, C.E., Malinen, J.T., Wakikawa, R., Nilsson, A., Tuominen, A.J.: Internet Connectivity for Mobile Ad Hoc Networks. Wireless Communications and Mobile Computing 2, 465–482 (2002)
16. Benzaid, M., Minet, P., Agha, K.A., Adjih, C., Allard, G.: Integration of Mobile-IP and OLSR for a Universal Mobility. Wireless Networks 10, 377–388 (2004)
17. Broch, J., Maltz, D.A., Johnson, D.B.: Supporting Hierarchy and Heterogeneous Interfaces in Multi-Hop Wireless Ad Hoc Networks. In: ISPAN 1999, pp. 370–375 (1999)
18. Ergen, M., Puri, A.: MEWLANA-mobile IP Enriched Wireless Local Area Network Architecture. In: Proceedings of VTC 2002-Fall, vol. 4, pp. 2449–2453 (2002)
19. Bernardos, C., Calderon, M.: Survey of IP Address Autoconfiguration Mechanisms for MANET. In draft-bernardos-manet-autoconf-survey-00.txt (2005)
20. Weniger, K., Zitterbart, M.: Address Autoconfiguration in Mobile Ad Hoc Networks: Current Approaches and Future Directions. IEEE Network (2004)
21. Mona, G., Philipp, H., Christian, P., Vasilis, F., Hamid, A.: Performance Analysis of Internet Gateway Discovery Protocols in Ad Hoc Networks. In: WCNC 2004, Atlanta, GA, USA, pp. 120–125 (2004)
22. Perkins, C., Belding-Royer, E., Das, S.: Ad hoc On-Demand Distance Vector (AODV) Routing. Internet rfc3561.txt (2003)

Connectivity Gateway Discovery in MANETs

Antonio J. Yuste[1], Alicia Triviño[2], Fco. David Trujillo[3], Eduardo Casilari[3],
and Antonio Díaz-Estrella[3]

[1] Department of Telecommunication Engineering
University of Jaén
Alfonso X El Sabio, 28
23700 Linares, Spain
ajyuste@ujaen.es

[2] Department of Lenguajes y Ciencias de la Computación
University of Málaga
Campus Universitario Teatinos, s/n
29071 Málaga Spain
atc@uma.es

[3] Department of Electronic Technology
University of Málaga
Campus Universitario Teatinos, s/n,
29071 Málaga Spain
{fdtrujillo,ecasilari,adiaz}@uma.es

Abstract. The integration of mobile ad hoc networks into IP-based access networks demands the presence of a gateway which is responsible for propagating some configuration parameters by means of Modified Router Advertisement (MRA) messages. This function may be accomplished on demand by the mobile nodes (reactively), periodically activated by the gateway (proactively) or by combining the two approaches. In the proactive mechanisms, the interval of emission of the MRA messages (T) may significantly affect the network performance. The optimum value for T depends on the network conditions such as the position of the sources, the mobility of nodes, etc. Therefore, an autonomous and adaptive technique to dynamically configure T is strongly recommended. In this sense, a new adaptive gateway discovery mechanism is proposed in this paper. The adaptation is achieved by a control system which has been conveniently configured by means of an analytical model. Simulation results show that the proposed adaptive mechanism improves the conventional proactive schemes.

Keywords: Ad-hoc networks, hybrid networks, Gateway interconection.

1 Introduction

The use of wireless access to connect to external networks such as the Internet is a key issue for the ubiquitous computing paradigm. Under some specific geographical constraints, the deployment of a wireless infrastructure that guarantees the mobile nodes acces within a particular area is economically unfeasible. In

L. Cerdà-Alabern (Ed.): Wireless and Mobility 2008, LNCS 5122, pp. 128–141, 2008.

these cases, the wireless networks could benefit from Mobile Ad Hoc NETwork (MANET) technology to reduce the costs related to the deployed equipment. Specifically, the use of multihop communications easily eliminates the existence of dead zones in wireless networks as well as it helps to extend the coverage area of GGSNs (Gateway GPRS Support Node) in the UMTS (Universal Mobile Telecommunications System). However, the use of MANET in this context prompts several technological challenges that need to be overcome. Concerning the architecture, the infrastructured networks offer an Access Router (or a GGSN in the UMTS), which provides the link to external hosts. In an IPv6 context, the Access Router periodically announces Router Advertisement (RA) messages which contain some configuration parameters such as the prefix information in order to allow mobile nodes to construct their own global IPv6 addresses [1]. These messages are generated with link-local addresses and therefore cannot be forwarded [2]. If only conventional Access Routers are used to connect MANET to the Internet, nodes placed outside the coverage area of the Access Routers will not obtain these configuration parameters and, consequently, will not be able to autoconfigure their own IPv6 addresses. This drawback is overcome by incorporating a gateway which is connected to the Access Router. Although the characteristics of the gateway differ in the proposed mechanisms, the most popular integration supports for Internet connectivity consider that the Access Router's firmware is customized to incorporate the gateway functionalities [3] [4]. In this sense, the gateway executes two main tasks. Firstly, it provides the ad hoc routing capabilities that are absent in conventional Access Routers so downlink traffic can be conveniently re-routed. On the other hand, the gateway generates Modified Router Advertisement (MRA) messages, which contain similar information to RA messages but which can be propagated in the MANET. As these messages are received by all the nodes in the MANET, they can autoconfigure their own IPv6 addresses to be globally reachable by any terminal in the Internet.

In the global connectivity support [3], the emission of MRA messages could be accomplished by three different strategies. As in the case of ad hoc route discovery, the Internet gateway could disseminate this message periodically, i.e. proactively. On the other hand, in the reactive scheme MRA messages are generated only on-demand when the Internet gateway receives a Modified Router Solicitation (MRS) message emitted by an ad hoc node requiring a route to the Internet. The MRS could also be generated by mobile devices when either the information related to proactive MRA messages or the route established by their reception become invalid. Finally, the hybrid approach combines the previous techniques. MRA messages are broadcast in an area close to the Internet Gateway while the devices located outside this zone demand the MRA information by generating MRS messages [5].

In this paper we will focus on the proactive gateway discovery. Under this scheme, when a mobile node receives the MRA message, it updates the route entry to the gateway and then, rebroadcasts the advertisement message. As all nodes update their tables, nodes do not require to wait for a response to an MRS message to start communicating with any external destination node. The parameter T or the frequency with which MRA messages are generated by the Internet gateway clearly affects the network performance. A low T value could unnecessarily consume the limited MANET resources such as power and bandwidth. On the other hand, a high T value could lead to the storage of stale routing information in the nodes so when they route packets to the gateway, packets losses may result and nodes are obliged to send an MRS message in order to get a correct route to the gateway. Thus, the advertisement period must be carefully selected in order to prevent the MANET from being gratuitously flooded with control packets (MRA or MRS messages) .

The aim of this paper is to present and evaluate an algorithm to automatically adapt the T parameter to the network conditions so that network performance is improved. Thus, the goal is to decrease control traffic without increasing loss nor delay of the user's packets. Our proposed algorithm is based on the estimation of the network connectivity from the percentage of nodes located in the transmission range of the gateway. In this sense, when there is a large number of neighbours, we assume that shorter routes are required for external communications. As shorter routes are usually associated to longer lifetimes, the routes could be updated less frequently, that is, the T could be set to a higher value [6]. The proposed tuning technique is supported by a control system which is configured by means of statistical properties regardless of the future location of the gateway or the topologies of the MANET.

The rest of this paper is organized as follows. In Section 2, we present the related work. Section 3 describes the proposed adaptive gateway discovery mechanism. A Simulation environment and results are presented in Section 4. Finally, conclusions are drawn in Section 5.

2   Related Work

The related work has been divided in two parts. The first part concerns some other adaptive algorithms presented in the related literature. The second section describes the model mobility patterns that will be employed in the simulations.

2.1   Adaptive Gateway Discovery Schemes

In gateway discovery mechanisms, the reception of MRA messages allows the stateless address configuration as well as the update of the routes to external hosts. In proactive schemes, the reception depends on the interval of emission of MRA messages (T). Additionally, in the hybrid gateway discovery mechanisms, the zone around the Gateway where these messages are proactively propagated, defined by the TTL (Time to Live) value, also determines which nodes receive the MRA messages. The optimum value of these two parameters clearly depends on the network conditions such as the position of the sources, the number of sources or the mobility conditions. Therefore, the TTL and the T values should be set according to the network conditions. As nodes in MANETs can move arbitrarily, the network conditions frequently vary so some algorithms have been proposed to dynamically set these two parameters.

Concerning the adjustment of the TTL value, one of the first adaptive algorithms was the Maximal Source Coverage (MSC)[7]. According to this proposal, the T is set to a fixed value. Meanwhile the gateway will send out the next advertisement message with the TTL equal to the minimum number of hops required to reach all of the sources that use this gateway to communicate with external hosts.

The optimization of the timing of MRA messages (set with the T parameter) was studied in [8]. In this approach, the authors suggest that the appropriateness of broadcasting an MRA message depends on the number of active sources that communicate with external hosts as well as the number of intermediate nodes that forward packets to the Internet gateway. With these two parameters, the authors define the so-called Regulated Mobility Degree (RMD). When this factor exceeds a pre-established threshold, the MRA message is sent. The main difficulty of this proposal is determining the threshold as it also depends on the network conditions.

In [9], the authors suggest the use of an auto-regressive filter to simultaneously adjust the T and its TTL value. To do so, the authors recommend monitoring the traffic load in Internet gateways so that the T and TTL can be conviently set using a feedback controller. The proposed tuning is based on the changes of link stability, the traffic rate and the number of received MRS. However, no specific formulation is presented and no evaluation is shown.

Another proposal to adapt the interval of emission of MRA messages is presented in [10]. In this proposal, the parameter T is dynamically tuned by means of estimating the number of reactive route solicitations that the nodes emit in a given time interval. With this purpose, the authors use an auto-regressive filter for the estimation. Although the proposal outperforms conventional proactive schemes in scenarios characterized by a low mobility, the benefits decrease when the network load increases and when the nodes speed increases.

In our proposal, the adjustment is based on the number of MRA messages retransmitted by the gateway neighbours. This parameter is assumed to be associated to the percentage of nodes that are close to the gateway and, therefore, have shorter routes for their external communications. Shorter routes are expected to have longer lifetimes [6] so their routes to the gateway do not need to be frequently updated. This indirect measurement of the network connectivity is employed to set the T parameter. So, if the network has a high connectivity, T can be increased. Initial results with this scheme were presented in [11].

2.2   Mobility Patterns

The performance evaluation of adaptive algorithms is usually carried out with just one type of scenario and one mobility pattern. The scenario is usually rectangular while the Random WayPoint Model (RWP) [12] is massively used as mobility pattern [10][11][13] . One of our goals was to verify if our proposed algorithm was able to reasonably adjust the T parameter under other mobility

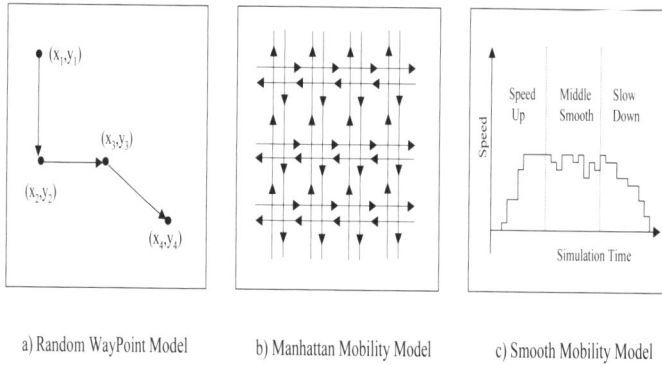

a) Random WayPoint Model   b) Manhattan Mobility Model   c) Smooth Mobility Model

Fig. 1. Mobility Patterns

conditions. With this purpose, in addition to the well-known RWP, the analysis of the proposed algorithm uses other two mobility models: the Manhattan Mobility Model [14] and the Smooth Mobillity Model [15]. The mobility patterns employed are briefly described in the following sections and some properties of them are depicted in Fig. 1.

Random Waypoint Model (RWP). The Random Waypoint model is commonly employed in the MANET research community. The implementation of this mobility model is as follows: at every instant, a node randomly chooses a destination and moves towards it with a speed chosen randomly (according to a uniform distribution) from $[1, V_{max}]$, where V_{max} is the maximum allowable velocity for every mobile node. When the destination is reached, the node stops for a duration defined by the 'pause time' parameter. After this duration, the node again chooses a new random destination within the area and repeats the whole process again until the simulation ends. A minimum and not null speed for the nodes movements is set according to [16].

Manhattan Mobility Model (MH). The Manhattan model emulates the movement of automobiles on streets defined by a rectangular grid-type map [14]. It can be useful in modeling movement in an urban area where a pervasive computing service between portable devices is provided. The grid is composed of a number of horizontal and vertical streets. The mobile node can move along the grid of horizontal and vertical streets on the map. At an intersection of a horizontal and a vertical street, the mobile node can turn left, right or go straight with certain probability. The Manhattan mobility model is also expected to have high spatial dependence and high temporal dependence as it imposes geographic restrictions on node mobility too.

For our simulations, in the grid there is a street every 100 meters both horizontally and vertically. The streets are two-way and the probability of turning at intersections is the same for all directions.

Smooth Mobility Model (SM). In [15] the authors define one movement as an entire motion from the time that a node starts to the moment when it stops moving. Based on the physical law of a smooth motion, a movement in the smooth model contains three consecutive moving phases: Speed Up phase, Middle Smooth phase, and Slow Down phase.

The parameters that we have employed in the different scenarios are the same as those defined in [15]: the time slot Δt of each step is set to 1 second, the memory parameter ζ is equal to 0.5, and the range of each moving phase duration time is set to [6, 30] seconds.

3   Proposal for Adaptive Gateway Discovery

In the proposals of proactive gateway discovery schemes, the interval of emission of MRA messages (T) is fixed to a constant value. However, as it has been remarked, the optimum value of T depends on the network conditions such as the load, the node mobility, the number of traffic sources, etc. This dependence is shown in Fig. 2 where the normalized routing overhead, that is, the number of control packets required to receive a data packet, is computed for different T values and for the scenarios A and B described in Table 1. Figure 2 shows that for low values of T, the MRA messages are gratuitously emitted as they are not employed to update the routes. On the other hand, when T is set to a high value, the mechanism behaves as a reactive gateway discovery scheme and, in some cases, the proactive gateway discovery outperforms the reactive ones. The graphic proves that there is an optimum value for T but it also illustrates that this value depends on the characteristics of the network. Specifically, for scenario A, the optimum T value corresponds to 7 seconds while for scenario B, the optimum T value is 17 seconds.

Figure 2 confirms the need for developing an algorithm that enables the gateway to dynamically set T to the optimum value. In our proposal, the gateway adjusts T taking into account the number of received MRA messages which are

Table 1. Simulation parameters

	Common Parameters
Simulation Area	1500 x 300 m^2
Mobility Pattern	Random WayPoint Model
Gateway	One in the center
Traffic	10 Constant Bit Rate sources
Rate	15 packet/s
	Scenario A
Maximum speed	5 m/s
Minimum speed	1 m/s
	Scenario B
Maximum speed	2 m/s
Minimum speed	1 m/s

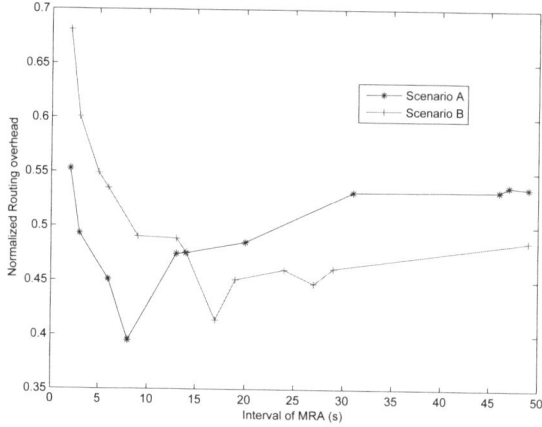

Fig. 2. Optimum T values for different Scenarios

retransmitted by its neighbours. This measurement provides an evidence of the network connectivity. In other words, when the gateway receives many MRA messages from its neighbours, it means that all these nodes have appropriately updated the routing entry to the gateway. Additionally, when the number of neighbours is high, we may assume that most routes to external hosts are composed of a low number of hops. This characteristic could lead to routes with a longer lifetime [6]. Under these circumstances, the routes do not need to be frequently updated so T could be incremented. On the other hand, when the gateway receives few MRA messages from its neighbours, it should decrease T to guarantee that nodes keep a valid route to the gateway when they require connection to the Internet. In order to determine the optimum T value as a function of the number of received MRA messages, we have employed the control system function shown in Fig. 3. The number of MRA messages received by the gateway is the input of this function while T is its output. The measurement of the number of received MRA messages will be carried out throughout T. The message count is carried out in a simple way: when the gateway sends an MRA message, it counts the number of copies of the recently generated MRA message during the next second. If no MRA message is received, the gateway sends the message again.

In this control system, we have adopted a linear relationship between the number of received MRA messages and the optimum T value. The decision to employ this function was based on the extensive simulations conducted, and the results of these simulations are presented in Fig. 4. The parameters of these simulations are similar to the Table 1. The plot shows the number of received MRA message versus the number of gateway neighbours. Although a percentage of the MRA messages is lost due to the interferences and collisions, the relationship between these variables can be considered linear.

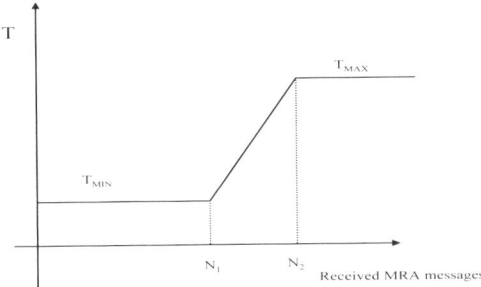

Fig. 3. Output function of the control system

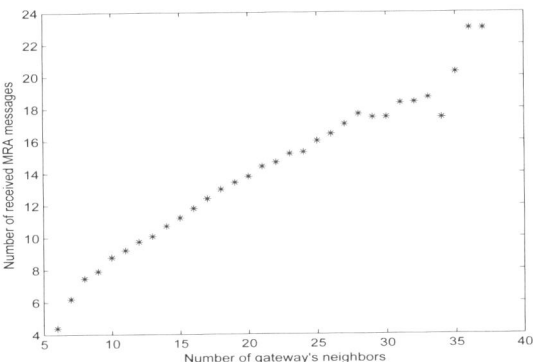

Fig. 4. The number of received MRA messages versus the number of gateway neighbours

In order to conveniently set the parameters N_1 and N_2 in the control system, the probability p that a node is located in the coverage area of the gateway is obtained. The gateway can always compute this probability regardless of the mobility model chosen. The probability may depend on a variety of factors including the number of mobile nodes, the topology or even the gateway location. Once the expression of p is obtained, the probability that there are n nodes in the coverage area can be computed as:

$$g(n) = \binom{N}{n} p^n (1 - p)^{N-n} \qquad (1)$$

It is a binomial distribution where N is the total number of mobile nodes in the MANET. In some integration supports such as the popular Global Connectivity [3], the gateway implements additional functionalities based on the registration that the mobile node performs when it enters the MANET. Under these circumstances, the gateway can easily know the number of nodes that form

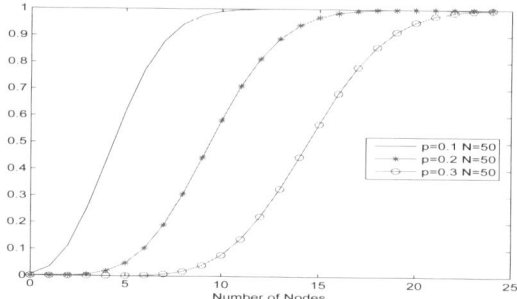

Fig. 5. Cumulative distribution function of the number of nodes that are the gateway neighbours

the MANET. If N is known, the p value can be obtained from the average value of the MRA messages received.

We can obtain the cumulative density function of $g(n)$, that is, $G(n)$ as shown in Fig. 5 with several p values. The values of N_1 and N_2 have been chosen from the linear zone of $G(n)$. N_2 will be set to the mean number of MRA messages received and N_1 the mean divided by 4. Thus the values of N_1 and N_2 are dynamic and change whenever an MRA message is sent by the gateway. By choosing N_2 as the average of the MRA messages received, we do not need to know the value of either N or p, and the algorithm adapts to changes in the number of active nodes or speed.

Concerning the limits of the T values, the standard value of 2 seconds has been chosen for T_{min} while the value of 20 seconds has been chosen for T_{max}. As shown in Fig. 2, when the T value outlasts 20 seconds, the normalized overhead is kept constant as the gateway discovery algorithm basically behaves as a reactive scheme.

4   Performance Evaluation

Due to the difficulties associated to real tests, the benefits of the adaptive gateway discovery have been verified by the use of simulations. In this sense, it was necessary to develop a software module that includes the algorithm in the Global Connectivity support [3]. This module has been integrated into the open source Network Simulator tool, ns-2.29 on a Linux machine [17]. Three different simulation settings, each one with a different mobility pattern, are defined and used in order to evaluate the performance of our proposal. In spite of this, there are several common parameters for the three simulations, presented in Table 2.

4.1   Scenarios

In order to evaluate the ability of the proposed algorithm to adjust the T parameter to the network conditions in a variety of scenarios, the simulations have been

Table 2. Common Simulation parameters

Transmision range	250 m
Ad hoc Protocol	AODV (Ad Hoc On Demand Distance Vector Routing)
	Local repair disabled
Link layer	Link layer detection enabled
	802.11 RTS/CTS enabled
Mobility pattern	Maximum speed:2 to 5 m/s
	Pause Time: 10 s
CBR	10 sources
	Rate: 15 packet/s

Table 3. Features of the Simulation Scenarios

Parameters	SCENARIO I	SCENARIO II	SCENARIO III
Simulation Area	1500 x 300 m^2	600 x 600 m^2	2500 x 500 m^2
Number of Nodes	50	75	100
Gateways Location	(750,150) m	(0,0)	(625,250) m
		(600,600) m	(1875,250) m

implemented in three different environments, each one with a different density of nodes, surface and gateway position. The characteristics are shown in Table 3:

- Scenario I. The first of the environments is a rectangular area with the gateway located in the center of the topology.
- Scenario II. The second environment is a square area. There are two gateways located in opposite corners of the square.
- Scenario III. The last one is wide rectangular area. There are also two gateways.

4.2 Results

In this section, we present simulation results based on our Adaptive Gateway algorithm (AGW), and compare them to the MSC and the RMD described in Section 2.1. RMD parameters are set equal to the reference [8].

The simulation time is 500 seconds, according with [18]. Since we are interested in studying the behavior of MANET in a steady state, the first 50 seconds of the simulation are considered a warm-up period and they are not computed in the analysis.

The algorithms are tested using the following metrics::

- Packet Loss Rate (PLR). Packet loss rate is defined as the ratio of the number of lost packets to the total number of packets transmitted by the sources.
- End-to-End delay (Delay). It represents the average time that the received packets take to reach the destination, i.e. the node in the external network, from their origins.

– The Normalized Routing Overhead (NRO). Defined as the total number of control packets divided by the total number of received packets. For this computation, each hop that a control packet makes is considered a new control packet.

These parameters offer an estimation of the network performance. Firstly, PLR and Delay are the two most important parameters from the users' point of view. On the other hand, the NRO is an important measurement in energy-limited devices as it provides an estimation of the battery consumption.

The results are summarized in the following tables. The values correspond to the average measurements computed from 26 simulations. The improvement has been measured as (2). This parameter gives us an idea about the efficiency of our algorithm, with its value over 5% in almost all cases.

$$Improvement = 100 \left[1 - \frac{NRO(AGW)}{min(NRO(AGW), NRO(MSC))} \right] \qquad (2)$$

Table 4 shows the results when the RWP model is employed. The worst delay is obtained for the scenario II due to the position of the gateway at the corners and because the nodes are prone to concentrate in the center in the RWP [19], which means a greater number of hops are required to reach the gateway. Our proposed scheme, the AGW, always obtains better results for the three metrics used.

Table 5 illustrates the results associated to the MH model. Now delays in scenarios I and II are similar, but the third one is related to the lowest Delay. The PLR and NRO decrease from the scenario I to III. Based on these three metrics, the AGW is again better.

Table 6 shows the results obtained with the SM model. The three metrics decrease from scenario I, with the highest values to the third scenario, where the worst performance is achieved. Again, AGW outperforms the other algorithms.

For all movements, the PLR is greater in the first scenario even if just one gateway exists. In this scenario, in terms of NRO the mobility pattern that

Table 4. Results with the RWP model

RWP		SCENARIO I Max Speed (m/s)		SCENARIO II Max Speed (m/s)		SCENARIO III Max Speed (m/s)	
Metric	T	2	5	2	5	2	5
Delay (s)	RMD	0.0738	0.0862	0.1745	0.3367	0.0309	0.0377
	MSC	0.0689	0.0858	0.1252	0.2820	0.0303	0.0346
	AGW	**0.0686**	**0.0800**	**0.1069**	**0.1756**	**0.0254**	**0.0274**
PLR	RMD	0.0507	0.0650	0.0179	0.0459	0.0096	0.0120
	MSC	0.0524	0.0629	0.0278	0.0529	0.0106	0.0129
	AGW	**0.0472**	**0.0862**	**0.0153**	**0.0243**	**0.0057**	**0.0075**
NRO	RMD	0.3852	0.5163	0.1659	0.2592	0.1427	0.2185
	MSC	0.3928	0.4827	0.1678	0.2645	0.1484	0.2252
	AGW	**0.3604**	**0.4539**	**0.1548**	**0.2458**	**0.1333**	**0.2017**
	Improv. (%)	**6.44**	**5.97**	**6.69**	**5.17**	**6.59**	**7.69**

Table 5. Results with the MH model

MH		SCENARIO I		SCENARIO II		SCENARIO III	
		Max Speed (m/s)		Max Speed (m/s)		Max Speed (m/s)	
Metric	T	2	5	2	5	2	5
Delay	RMD	0.0782	0.1547	0.0725	0.0860	0.0216	0.0329
(s)	MSC	0.0763	0.1538	0.0659	0.0797	0.0211	0.0316
	AGW	**0.0695**	**0.1535**	**0.0543**	**0.0701**	**0.0194**	**0.0274**
PLR	RMD	0.0627	0.0903	0.0137	0.0216	0.0022	0.0053
	MSC	0.0632	0.0908	0.0142	0.0216	0.0023	0.0056
	AGW	**0.0571**	**0.0864**	**0.0111**	**0.0201**	**0.0019**	**0.0047**
NRO	RMD	0.27	0.6180	0.1435	0.2062	0.0681	0.1204
	MSC	0.3	0.6293	0.1527	0.2095	0.0736	0.1268
	AGW	**0.2461**	**0.579**	**0.1355**	**0.198**	**0.0615**	**0.1143**
	Improv. (%)	**8.85**	**7.99**	**5.57**	**3.98**	**9.69**	**5.07**

Table 6. Results with the SM model

SM		SCENARIO I		SCENARIO II		SCENARIO III	
		Max Speed (m/s)		Max Speed (m/s)		Max Speed (m/s)	
Metric	T	2	5	2	5	2	5
Delay	RMD	0.1523	0.2350	0.0861	0.1603	0.0230	0.0247
(s)	MSC	0.1485	0.2308	0.0822	0.1381	0.0200	0.0238
	AGW	**0.1464**	**0.2307**	**0.0588**	**0.0983**	**0.0197**	**0.0231**
PLR	RMD	0.0857	0.1288	0.0160	0.0368	0.0013	0.0018
	MSC	0.0862	0.1289	0.0167	0.0349	0.0018	0.0035
	AGW	**0.0827**	**0.1289**	0.0122	0.0283	**0.0010**	**0.0016**
NRO	RMD	0.6121	0.8503	0.1155	0.1885	0.0476	0.0781
	MSC	0.6132	0.8503	0.1285	0.1959	0.0904	0.1429
	AGW	**0.5761**	**0.7975**	**0.1089**	**0.18**	**0.0417**	**0.073**
	Improv. (%)	**5.88**	**6.21**	**5.71**	**4.51**	**12.39**	**6.53**

offers the best features is the MH while the SM provides the worst performance. For scenarios II and III, NRO is the best when the SM is used to model node mobility.

The comparison between the RMD and MSC algorithms illustrates that the RMD scheme is better regarding PLR and NRO, but not in the Delay metric. However, Delay, PLR and the NRO are always improved when the proposed adaptive discovery scheme is employed as shown in Table 4 , Table 5 and Table 6.

5 Conclusions

This paper presents a new method to optimize the process that enables Internet connectivity in multi-hop ad hoc networks. The optimization minimizes the control overhead by dynamically tuning the interval of emission of MRA messages (T) as a function of network connectivity. For this purpose, the algorithm estimates the network connectivity by means of the number of MRA messages received. The algorithm outperforms other schemes regardless of the node speed, the location of the gateways and the mobility patterns. The reduction of the overhead is a very desirable characteristic in mobile ad hoc networks because of node limitations regarding battery resources.

Acknowledgements. This work has been partially supported by the Ministerio de Ciencia y Tecnología, project No. TEC2006-12211-C02-01/TCM.

References

1. Narten, T., Nordmark, E., Simpson, W.: Neighbor Discovery for IP version 6. IETF RFC 2461 (December 1998)
2. Huitema, C.: IPv6: the new Internet Protocol. Prentice-Hall, Upper Saddle River (1998)
3. Wakikawa, R., Malinen, J.T., Perkins, C.E., Nilsson, A., Tuominen, A.J.: Global Connectivity for IPv6 Mobile Ad Hoc Networks. Internet Engineering Task Force, Internet Draft (work in progress)
4. Jelger, C., Noel, T., Frey, A.: Gateway and Address Autoconfiguration for IPv6 Ad Hoc Networks. Internet-Draft draft-jelger-manet-gateway-autoconf-v6-02.txt (April 2004)
5. Ratanchandani, P., Kravets, R.: A Hybrid Approach to Internet Connectivity for Mobile Ad Hoc Networks. IEEE Wireless Communications and Networking 3, 1525–1527 (2003)
6. Triviño-Cabrera, A., García-de-la-Nava, J., Casilari, E., Gonzãlez-Cañete, F.J.: An Analytical Model to Estimate Path-Duration in MANETs. In: 9th ACM/IEEE International Symposium on Modeling, Analysis and Simulation of Wireless and Mobile Systems (MSWiM 2006), Torremolinos, Spain (October 2006)
7. Ruiz, P.M., Gomez-Skarmeta, A.F.: Maximal Source Coverage Adaptive Gateway Discovery for Hybrid Ad Hoc Networks. In: Nikolaidis, I., Barbeau, M., Kranakis, E. (eds.) ADHOC-NOW 2004. LNCS, vol. 3158, pp. 28–41. Springer, Heidelberg (2004)
8. Rakeshkumar, V., Misra, M.: An Efficient Mechanism for Connecting MANET and Internet through Complete Adaptive Gateway Discovery. In: First International Conference on Communication System Software and Middleware (COMSWARE 2006), New Delhi, India, January 2006, pp. 1–5 (2006)
9. Ghassemian, M., Friderikos, V., Aghvami, A.: A generic algorithm to improve the performance of proactive ad hoc mechanisms. In: Sixth IEEE International Symposium on a World of Wireless Mobile and Multimedia Networks (WoWMoM 2005), Taormina, Italy, June 2005, pp. 362–367 (2005)
10. Triviño-Cabrera, A., Ruiz-Villalobos, B., Casilari, E.: Adaptive Gateway Discovery in Hybrid MANETs. In: Workshop on Applications and Services in Wireless Networks (ASWN 2007), Santander, Spain, May 2007, pp. 9–14 (2007)
11. Yuste, A.J., Trujillo, F.D., Triviño, A., Casilari, E.: An adaptive gateway discovery for mobile ad hoc networks. In: 5th ACM international workshop on Mobility management and wireless access (MOBIWAC), Chania, Crete Island, Greece, October 2007, pp. 159–162 (2007)

12. Bettstetter, C., Resta, G., Santi, P.: The node distribution of the random way-point mobility model for wireless ad hoc networks. IEEE Transactions on Mobile Computing, 257–269 (July-September 2003)

13. Ruiz, P.M., Gomez Skarmeta, A.F.: Enhanced Internet Connectivity for Hybrid Ad hoc networks Through Adaptive Gateway Discovery. In: 29th Annual IEEE Conference on Local Computer Networks (LCN 2004), Tampa, FL (November 2004)

14. Bai, F., Sadagopan, N., Helmy, A.: IMPORTANT: a framework to systematically analyze the Impact of Mobility on Performance of Routing Protocols for Adhoc Networks. In: Twenty-Second Annual Joint Conference of the IEEE Computer and Communications Societies. INFOCOM 2003, March 30- April 3, pp. 825–835. IEEE, Los Alamitos (2003)

15. Zhao, M., Wang, W.: Design and Applications of A Smooth Mobility Model for Mobile Ad Hoc Networks. In: Military Communications Conference. MILCOM 2006, October 23-25, 2006, pp. 1–7 (2006)

16. Yoon, J., Liu, M., Noble, B.: Random waypoint considered harmful. In: Twenty-Second Annual Joint Conference of the IEEE Computer and Communications Societies. INFOCOM 2003, March 30- April 3, pp. 1312–1321. IEEE, Los Alamitos (2003)

17. Network simulator, `http://www.isi.edu/nsnam/ns`

18. Casilari, E., Triviño-Cabrera, A.: A practical study of the Random Waypoint mobility model in simulations of ad hoc networks. In: 19th International Teletraffic Congress (ITC'19), Pekin, China, September 2005, pp. 115–124 (2005)

19. Bettstetter, C., Hartenstein, H., Pérez Costa, X.: Stochastic Properties of the Random Waypoint Mobility Model. Wireless Networks 10(5), 555–567 (2004)

Efficiency of Search Methods in Dynamic Wireless Networks

Gerhard Haßlinger[1] and Thomas Kunz[2]

[1] T-Systems Enterprise Services,
Deutsche-Telekom-Allee 7,
D-64295 Darmstadt, Germany
gerhard.hasslinger@telekom.de
[2] Carleton University,
Department of Systems and Computer Engineering,
1125 Colonel By Drive,
Ottawa, Ontario, Canada, K1S 5B6
tkunz@sce.carleton.ca

Abstract. Search methods in dynamic networks usually cannot rely on a stable topology from which shortest or otherwise optimized paths through the network are derived. When no reliable search indices or routing tables are provided, other methods like flooding or random walks have to be considered to explore the network. These approaches can exploit partially available information on network paths, but the search effort naturally increases with the lack of precise paths due to network dynamics. The problem is especially relevant for wireless technology with strict limitation on power consumption.

We compare the efficiency of random walks and flooding for exploring networks of small to medium size. Several scenarios are considered including partial path information support for search. Transient analysis and a bound are applied in order to evaluate the messaging overhead.

Keywords: search, routing, wireless networks.

1 Introduction

Exploration and search methods are generally required to enable services and content retrieval in communication networks. Even in fixed network areas of the Internet, where the topology is stable enough to establish standard routing protocols and search engines to locate nodes and information on them, they have to cope with continuous changes. More dynamic network structures are often encountered on peer-to-peer (P2P) overlays [13] [24] [25] as well as in multihop wireless networks such as wireless sensor or (mobile) ad hoc networks.

A search may refer to users, network nodes, information, content or services of any kind residing on network resources based on identifiers like IP addresses or hash values used in P2P networks. Although a single node is often addressed by a search, this can be extended to a set of nodes denoted as the target node set, each of which is

L. Cerdà-Alabern (Ed.): Wireless and Mobility 2008, LNCS 5122, pp. 142–156, 2008.
© Springer-Verlag Berlin Heidelberg 2008

able to respond successfully. Other cases, where several nodes have to be involved to get a result in a production chain or a distributed scheme, are not considered here.

Random walks proved to be a promising alternative for search in unstructured networks, especially in large scale networks [1] [7] [10] [13] [20] [25], where the standard method by flooding may be prohibitive, even with a limited hop count as in the first version of the Gnutella P2P network [24].

In this study, we investigate networks whose nodes are communicating wireless by sending broadcast messages into their neighborhood. We compare flooding and random walks with regard to underlying structures of network connectivity for grids and for cases of nodes being randomly spread over an area. With application to sensor and ad hoc networks, alternatives for routing and search methods have been discussed in recent work [2] [3] [19] [26], where basic random walks and flooding are often subject to inefficiency and high overhead.

For mobile ad hoc networks (MANETs), the dynamic nature of the network topology challenges routing protocols in new ways. For example, it has long been known that node mobility causes unicast routing protocols to perform poorly, as shown in [3] [9] [16] [22]. In the case of multicast routing, [17] similarly shows that routing protocol performance suffers with an increase in node mobility. Approaches such as simply flooding all packets in the network can be surprisingly competitive both in the resulting protocol performance and the induced overheads. But even in the absence of node mobility, multihop wireless networks experience highly dynamic topologies, challenging any routing protocol. For example, measurements in existing wireless mesh testbeds show that even static wireless links are highly asymmetrical and have time-varying behaviour due to interference, requiring new routing solutions, see for example [8].

We similarly showed in [23] how more accurately accounting for the physical realities of a wireless channel leads to poor routing performance of typical shortest-path routing protocols. In the case of Dynamic Source Routing (DSR) [15], one of the MANET routing protocols standardized by the Internet Engineering Task Force (IETF) [14], a packet delivery ratio of as low as 4% was observed, i.e., only one out of 25 data packets transmitted by a sender was successfully delivered to the intended receiver. One of the main reasons for this poor performance is the fact that shortest-path routing protocols tend to select relatively long-distance hops, which are then subject to interference, similar to the observations in [8]. Even if the routing metric is changed to select more stable links, the problems of sharing a limited bandwidth wireless channel, interference from transmissions in the neighborhood, etc. can still result in relatively poor overall performance. For example, collecting state information to improve routing protocol performance in particular for QoS routing in a static network with stable links has been shown to result in highly inaccurate views of the network state [18].

Almost all MANET routing protocols provide parameters to adapt the protocol behavior to the specific characteristics of the network in which they are deployed. For example, protocols that discover neighbors through periodic HELLO messages typically define the periodicity of these messages as a protocol parameter. This way, a larger HELLO interval sending fewer overhead messages can be chosen for relatively static networks. When a node's neighborhood changes at a more rapid pace, a smaller HELLO interval allows a node to learn about these changes faster, albeit at the cost of higher protocol overheads. We have done extensive evaluations of the Optimized Link State Routing (OLSR) protocol [6] and found that tuning these parameters has little influence on the protocol performance, though carefully choosing the appropriate parameter values can reduce protocol overheads [27]. Also, adjusting these control parameters does not increase the accuracy of any state information collected by individual nodes [18].

The IETF has recognized as well that MANET routing protocols such as OLSR and DSR may not meet the routing requirements of low power and lossy networks, chartering a new working group in early 2008 [14]. Our experience, reported in [17] shows that broadcasting/flooding can be superior to routing, as mentioned above. Furthermore, random walks have been shown to be beneficial in large unstructured networks, compared to flooding. In addition, it is demonstrated [2] that random walks essentially benefit even from imprecise and only partially valid information in support of a search or when many nodes in the network are able to respond, i.e. when it is sufficient to reach one node of a larger set.

In the sequel, we briefly compare properties of random walks and flooding using the transient analysis technique in section 2. In section 3, biased random walks in grid networks are evaluated by transient analysis from a more general view. A performance bound for this case is obtained by focusing on the distance to the target leading to a linear chain approach in section 4. Section 5 studies networks of uniformly distributed nodes as another basic modeling case in order to round up the evaluation followed by the conclusions.

2   Analysis of the Efficiency of Random Walks and Flooding

We denote the topology of a communication network as a graph $G = (V, E)$ with sets V of nodes and E of edges, which are assumed to be undirected. As main characteristics relevant to this study we consider

- the degree $d(a) = |\{k \mid k \in V, (a, k) \in E\}|$ of a node $a \in V$, i.e. the number of edges attached to the node with minimum $d_{min} = \min\{d(a) \mid a \in V\}$, and
- the hop distance h_{ab} between a pair of nodes a, b as the length of a shortest path from a to b, whose maximum is the network diameter $dia = \max\{h_{ab} \mid a,b \in V\}$.

In 2-dimensional grid networks as one of the evaluation examples, a node with coordinates $(x, y) \in V$ is connected to 4 neighbors $(x–1, y)$, $(x+1, y)$, $(x, y–1)$ and $(x, y+1)$ except for missing neighbors beyond the network boundaries at $x = 0$, $x = N$, $y = 0$, and $y = M$ (here N and M denote the grid size).

We follow a random walk through the network as a stepwise process, which proceeds from a node to a neighbor at the next hop. A random walk R of length L is denoted by the series $r_0, r_1, r_2, ..., r_L$ of visited nodes, where an edge $(r_{k-1}, r_k) \in E$ is chosen for the k-th hop ($1 \leq k \leq L$). Usually a random walk chooses its next hop with the same probability among all neighbors of the currently visited node

$$\forall\, a, k;\ (a, k) \in E: \quad p_{ak} = \Pr(r_{n+1} = k \mid r_n = a) = 1 / d(a). \tag{1}$$

The corresponding transition matrix $P = (p_{ak})$ determines a random walk as a Markov process, where the network nodes directly correspond to the states of the underlying Markov chain and edges to allowable transitions from a state to another.

2.1 Transient Analysis of Random Walks

When the exploration of a network by random walks of predefined length is evaluated using simulation [7] [10] [11] [20] [26], the results are subject to confidence levels, with long simulation runs being required to achieve tight confidence intervals. General analytical bounds for the convergence are valuable to ensure the principle behaviour, but often are not tight [7]. We follow an alternative classical approach using transient analysis [2] [13], which stepwise determines the probabilities $p_m^{(R)}(a)$ of a random walk R to enter a network node a at its m-th hop. When the random walk starts at a specific node s, then we have

$$p_0^{(R)}(s) = 1 \quad \text{and} \quad \forall a \neq s: \ p_0^{(R)}(a) = 0$$

as the initial distribution. In general, any arbitrary initial distribution can be considered. The transient analysis iteratively computes the distribution of the next hop location. From knowledge of $p_m^{(R)}(a)$, the distribution $p_{m+1}^{(R)}(a)$ for the next step is computed by

$$p_{m+1}^{(R)}(a) = \sum_{k:\,(k,\,a)\in E} p_m^{(R)}(k)\, p_{ka} \,. \tag{2}$$

For the purpose of network search performance, the probability $q_m^{(R)}(t)$ that a node t has been reached during a random walk of length m is the most important measure

$$q_m^{(R)}(t) = \Pr(\,\exists j;\, 0 \leq j \leq m:\, r_j = t\,).$$

The probabilities $q_m^{(R)}(t)$ are computed by the same transient step-by-step approach. The only modification required is to introduce an absorbing state at node t, which accumulates the probabilities for visiting that node during the walk. Therefore the transition equations (1) need to be changed only for departures from node t

$$\forall\, a, k;\, (a, k) \in E:\ q_{ak} = 1/d(a) \qquad \text{if } a \neq t;$$
$$\forall\, k;\, (t, k) \in E:\ q_{tk} = 0 \text{ if } k \neq t; \qquad q_{tt} = 1. \tag{3}$$

$q_m^{(R)}(t)$ is determined by applying equation (2) to the modified transition matrix

$$q_{m+1}^{(R)}(a) = \sum_{k:\,(k,\,a)\in E} q_m^{(R)}(k)\, q_{ka} \,. \tag{4}$$

The random walk defined by equations (3) and (4) has unchanged behaviour at all nodes except for t. Once it has reached t it stays there for ever. Consequently, $q_m^{(R)}(t)$ is monotonously increasing with m. Provided that the graph is irreducible and finite, node t will be reached sooner or later with probability 1 even in periodical cases

$$q_m^{(R)}(t) \leq q_{m+1}^{(R)}(t); \quad \lim_{m\to\infty} q_m^{(R)}(t) = 1.$$

The computational per hop complexity of the transient analysis due to equations (2-4) is proportional to the number $|E|$ of edges in the network. Therefore the effort to compute $p_m^{(R)}(a)$ is of the order $O(m|E|)$, which makes transient analysis applicable to large scale networks with millions of nodes.

2.2 Multiple Random Walks in Parallel

Random walks often can reduce the communication overhead, but they traverse the hops sequentially and thus usually spend much more time than flooding. Multiple random walks in parallel may be applied in a compromise between demands for low delay and low overhead. If a random walk is assured a success rate of $\sigma < 1$ within m steps, then k random walks of the same type in parallel each with m steps reduce the failure rate from $1 - \sigma$ to $(1 - \sigma)^k$. Thus a success rate of $\sigma = 90\%$ is improving up to 99.999% when 5 random walkers are combined in parallel. A single walk often achieves this success rate in less than mk steps but even then needs up to k-fold time.

2.3 Flooding

In the simplest case, flooding will spread a request from a node to all its neighbors which repeat it in order to contribute to an exhaustive flood covering the entire network. When the same request is received several times at a node from different neighbors, then only the first receipt is forwarded and later ones are discarded. Wireless networks spread messages through broadcasting over a limited range. Therefore only one message is required per node for flooding, whereas $d(a)$ messages are sent per node in overlay or meshed point-to-point networks. Therefore flooding is more competitive in broadcast environments.

In order to reduce the messaging overhead, flooding is usually restricted by a predefined hop limit h [24]. An appropriate value of h may be derived from knowledge of the network structure and a demand for coverage. Without that knowledge, h may be initialized with a small value and stepwise increased if the search radius turns out to be insufficient. In a 2-dimensional grid there are a maximum of $2h^2 + 2h + 1$ nodes within a distance of $\leq h$ hops, if the surrounding of the node is not further restricted by the grid boundary.

3 Application to Planar Grid Networks

In order to study the tradeoffs in the performance of flooding and random walks, we consider large 2-dimensional grids where the start and target node are located close to the center. The transient analysis approach can easily cope with grid sizes of several hundred in both dimensions. The results are valid for infinite planar grids in the sense that we extend the size until we are sure that the grid boundary has no impact, i.e. any boundary is touched by the biased walk only with negligible probability, e.g. $< 10^{-10}$. Note that infinite grids constitute a scenario of deterministic and worst possible performance, since in case of reaching a boundary the walk is always reflected in the direction towards the destination in the center, avoiding detours outside the boundary. For unbiased random walks on infinite grid networks, the mean recurrence time to the starting point is known to be infinite as well as the mean time to find a target even in the near of the origin.

The search for a single target node is carried out by a biased random walk [2] utilizing partial routing information available in the network. This model assumes that a node is able to forward an incoming search request with probability ρ $(0 < \rho < 1)$ into the direction downstream to the target. With probability $1 - \rho$ the node has no valid information in support of the search due to network dynamics, e.g. mobility or churn of nodes. Then the forwarding is done at random.

Starting from a grid node (a, b) with distance $\Delta = |a - x| + |b - y|$ to the target (x, y), we have to distinguish whether (a, b) is in line with the target, such that $a = x$ or $b = y$. If they are in line, then one of the 4 neighbors of (a, b) is closer to the target while the three other neighbors have distance $\Delta+1$. The biased random walk chooses each of those three with probability $(1-\rho)/4$ and the closer one with probability $(1 + 3\rho)/4$. Otherwise, if they are not in line then there exist two neighbors at distance $\Delta-1$ and two at $\Delta+1$ as depicted in Fig. 4. Both closer neighbors are then chosen with probability $(1+\rho)/4$ each and both ones at larger difference with probability $(1-\rho)/4$, respectively.

The model obviously does not cover all realistic scenarios, since it is left open how a node decides whether it has valid up-to-date information to direct the next step to the target. In addition, a homogeneous information distribution is implied, whereas network structures are often inhomogeneous, e.g. hierarchical, and information is usually more precise near the target.

Fig. 1 shows how many steps are required for a biased random walk to reach the target at a distance of 1, 2, … 20 hops with a success rate of 90%. Start and target

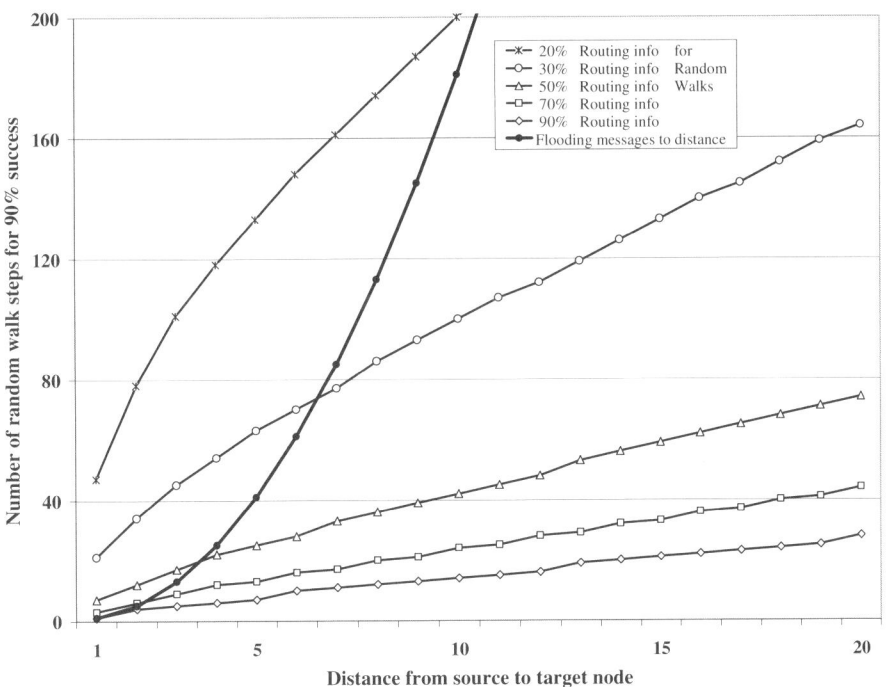

Fig. 1. Overhead of biased random walk search for 90% success

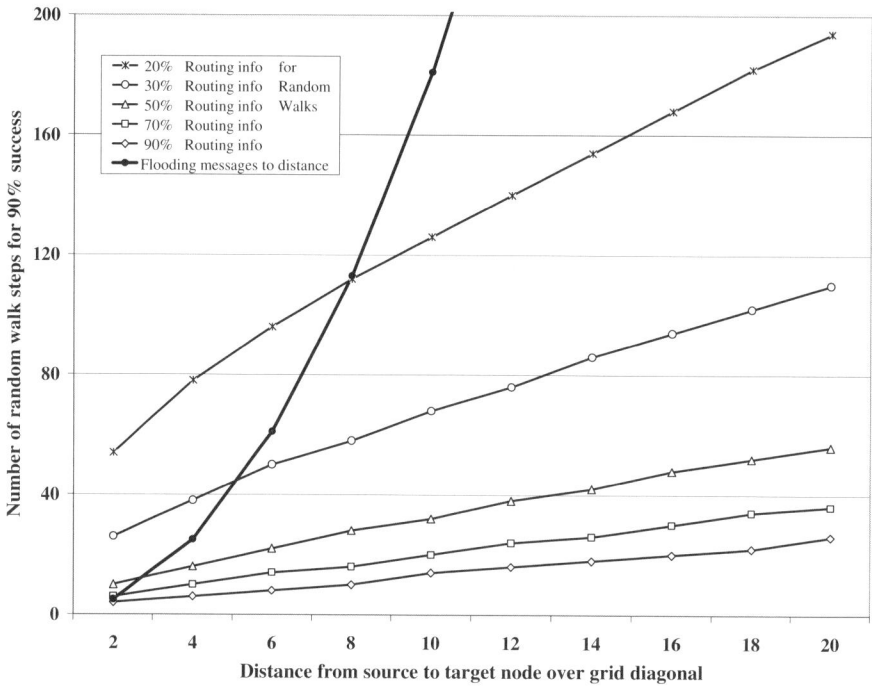

Fig. 2. Effort of random walk with target reached over grid diagonal

node are assumed to be located on a common line in the grid. The 3 lower curves with 50% or more of the nodes having valid routing information show that random walks are then much more efficient than flooding, which is characterized by the quadratic curve $2\Delta^2 - 2\Delta + 1$ for the number of nodes within distance $\Delta - 1$. Since the distance Δ is not known a priori, more effort may be necessary to iteratively find an appropriate hop count limit.

While the analysis in Fig. 1 is for targets located on a common line with the source, we next consider the opposite cases of targets to be reached over a diagonal in the grid. Fig. 2 shows the corresponding results for hop distances up to 20 and the same parameters, revealing an even better performance of random walks. The improvement is plausible from the fact, that if the current node and the target are on a line, only one of the four neighbors is closer to the target while otherwise two of the four neighbors lead in the direction to the target.

Fig. 3 gives the corresponding results for a success rate of 99% instead of 90% used in the previous cases. For small distances, a single random walk often needs more hops than two independent random walks together for the same success rate, which makes multiple walks preferable. For larger distances, a single walk reaches the 99% success at less than twice the number of hops required for 90% success.

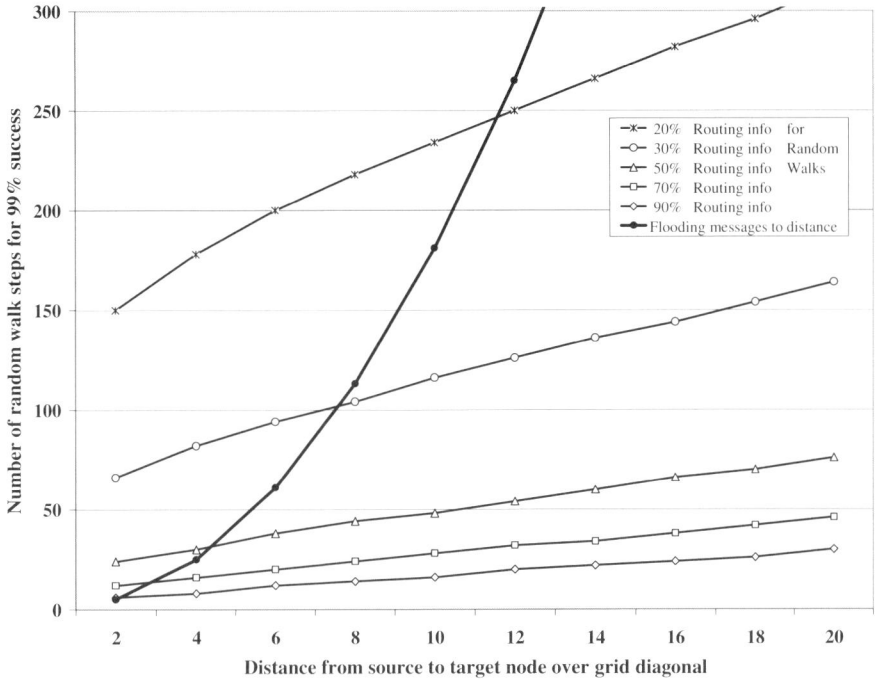

Fig. 3. Overhead of biased random walk search for 99% success

4   A Bound on Biased Random Walk Performance

Assuming that a biased random walk decrements the distance Δ to the target in each step with probability q or increments the distance with probability $p < q$ or stays at the same distance with probability $1-p-q$, the process of approaching the target becomes a simple birth-death process, as is well known e.g. for M/M/1 queueing systems.

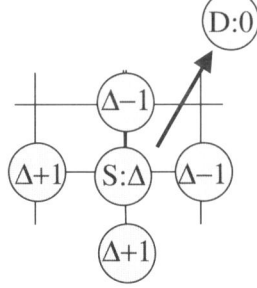

Fig. 4. Hop distances to a destination in a grid

The reduction to a 1-dimensional view is appropriate when the random walk has the same behaviour at each node and when no boundaries have to be regarded in a grid or other network structure. The same simplification is applied for modeling user mobility [21] and is analyzed via the corresponding diffusion process for packet delivery in sensor networks [5].

The transient behaviour of the birth-and-death process is characterized by a geometrically distributed number of hops until the next change in the distance with parameter $1 - p - q$. For 2-dimensional grids, the distance changes in each step, i.e. $p + q = 1$. Starting at a distance Δ from the target, the distance Δ_m after m steps has a binomial distribution, if m only counts steps with changes in the distance

$$\Pr\{\Delta_m = \Delta + m - 2k\} = \binom{m}{k} \omega^k (1 - \omega)^{m-k} \quad \text{for } k = 0, \ldots, m \text{ where } \omega = q/(p+q).$$

In principle, negative distances are also included, which are only reachable by previously traversing the target at distance 0. The complete probability mass below 0 indicates that the search was already successful, i.e. state 0 should be an absorbing bound. When we continue the analysis of the non-truncated birth-death chain, the mean $E(\Delta_m)$ of Δ_m is given by

$$E(\Delta_m) = \Delta - (2\omega - 1) m \quad \text{for} \quad E(\Delta_m) \leq 0.$$

For $m \geq \lceil \Delta/(2\omega - 1) \rceil$ we can conclude that the target has been reached at least with 50% probability, since a binomial distribution is symmetrical and has most of its probability mass in the negative part when the mean is negative, see Fig. 5.

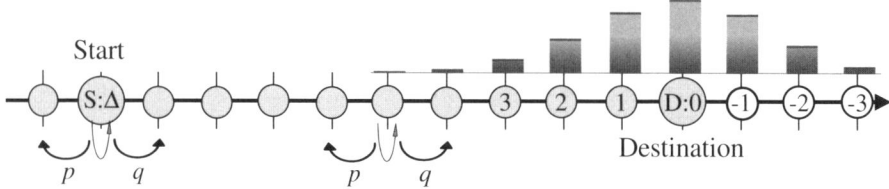

Fig. 5. Linear Markov chain approach for the distance to the target

This gives a clear and simple hint on the number m of steps required for a successful biased random walk. For high success probabilities $1 - 10^{-k}$ we can include the variance $\sigma^2(\Delta_m)$ of the distribution in the analysis via Chebychev's bound

$$\Pr\{\Delta_m > E(\Delta_m) + 10^{k/2}\sigma(\Delta_m)\} \leq 10^{-k}.$$

The mean $E(\Delta_m) = \Delta - (2\omega - 1)m$ and variance $\sigma^2(\Delta_m) = 4\,\omega(1 - \omega)m$ of the binomial distribution leads to a bound on $\Pr\{\Delta_m > 0\}$, from which we derive a sufficient number m_{Che} of steps in order to reach the destination with probability $1 - 10^{-k}$:

$$m_{Che} = \left\lceil \frac{\Delta}{2\omega - 1} + \frac{2\omega(1 - \omega)10^k}{(2\omega - 1)^2} \left(1 + \sqrt{1 + \frac{2\omega - 1}{\omega(1 - \omega)} \frac{\Delta}{10^k}}\right) \right\rceil. \tag{5}$$

The result shows that the number of required steps can be bounded by a linear function of the distance Δ, where a constant term is involved depending on the success probability, which becomes very large with increasing k. In this way, a linear behaviour is confirmed in general, although the bound is often far from being tight. The linear factor $1/(2\omega - 1)$ in the first term of equation (5) is confirmed as the asymptotical behaviour for all curves of Fig. 6.

The success probability can also be calculated directly from the tail probabilities of the binomial distribution, which yields much more precise results on the expense of a more complex but still tractable calculation.

In order to apply the bound to random walks in a grid we obtain

$$q = \Pr\{\Delta_{m+1} = \Delta_m - 1\} = (2+2\rho)/4 \quad \text{and} \quad p = 1 - q = \Pr\{\Delta_{m+1} = \Delta_m + 1\} = (2-2\rho)/4$$

if the current node is not in a line with the target or otherwise $q = (1+3\rho)/4$; $p = (3 - 3\rho)/4$. We ignore the latter case for nodes in line with the target and adopt $q = (2 + 2\rho)/4$, $p = (2 - 2\rho)/4$. In this way, we get an approximation rather than a bound, which is valid for most of the traversed nodes except for those on both grid lines crossing at the target. Fig. 6 compares the evaluations with the transient analyses of Fig. 2 to the derivation from the bound, confirming a good overall match.

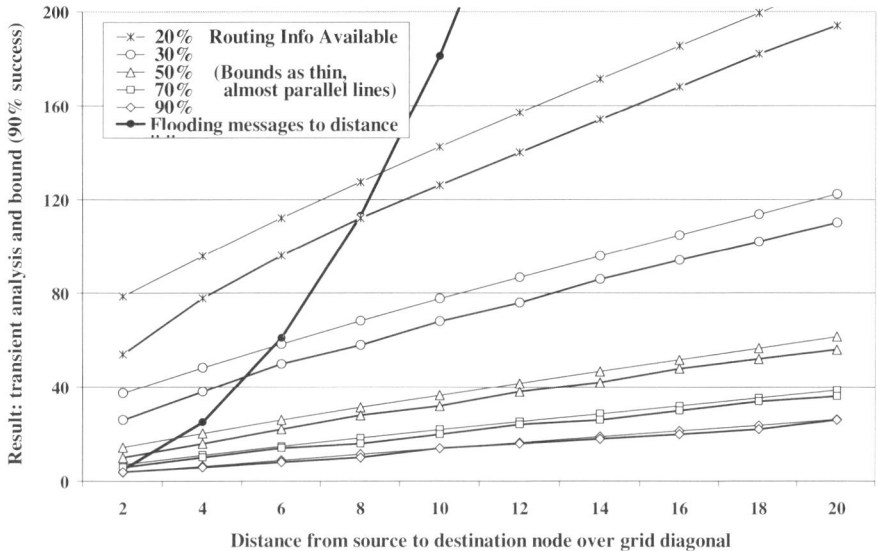

Fig. 6. Applying the bound to the examples of Fig. 2

5 Application to Uniformly Distributed Nodes in Wireless Networks

As another scenario, we consider a network with nodes being uniformly distributed over an area. Mobility models and sensor networks often exhibit similar random node locations close to a uniform distribution. In this section, we place a predefined number N of nodes in a square of size S^2, where each node is assumed to be able to send broadcast messages within a radius R ($R < S$) such that it has a direct one hop connection to all other nodes located at distance less than R.

Again we assume that up-to-date routing information is only partial, impaired by dynamics in the network caused by mobility or unreliable links or nodes. Again we assume a node to be able to select a neighbor closer to the destination with probability ρ ($0 < \rho < 1$) or otherwise to pick a random neighbor who next becomes responsible to forward the message.

The considered network topology is characterized by two parameters N and R/S. In the following examples we at first use a random number generator to choose N node locations in a square, and then compute the adjacency matrix $A = (a_{ij})$ indicating the directly reachable neighbors of a node by checking if the distance between two nodes is less than the reachability radius R. Based on A, we determine the hop distance matrix $H = (h_{ij})$ between each pair of nodes and further characteristics of the network topology, i.e. the network diameter $dia = \max(h_{ij})$ and the mean hop distance between nodes, as well as the connectivity degree of the nodes.

Finally, we consider biased random walks in the given network topology using transient analysis to evaluate their performance. Therefore we randomly select source-destination node pairs with hop distances ranging from 1 to the network diameter. For the cases studied in Fig. 7 - Fig. 10, we computed the average number of steps required for 99% success including up to 10 source-destination pairs for each distance or fewer pairs if 10 are not encountered at this distance. Table 1 summarizes the main characteristics of the considered network topologies.

Table 1. Parameters of the networks under study in Fig. 7 - Fig. 10

Network Scenario	Case of Fig. 7	Case of Fig. 8	Case of Fig. 9	Case of Fig. 10
Number of nodes N	200	200	1000	1000
Transmit radius (R/S)	0.016	0.03	0.003	0.01
Network diameter dia	18	9	34	16
Mean hop distance	7.0	4.2	13.3	6.3
Minimum node degree	1	6	1	6
Mean node degree	9.1	15.6	8.8	28.9
Maximum node degree	17	27	18	48

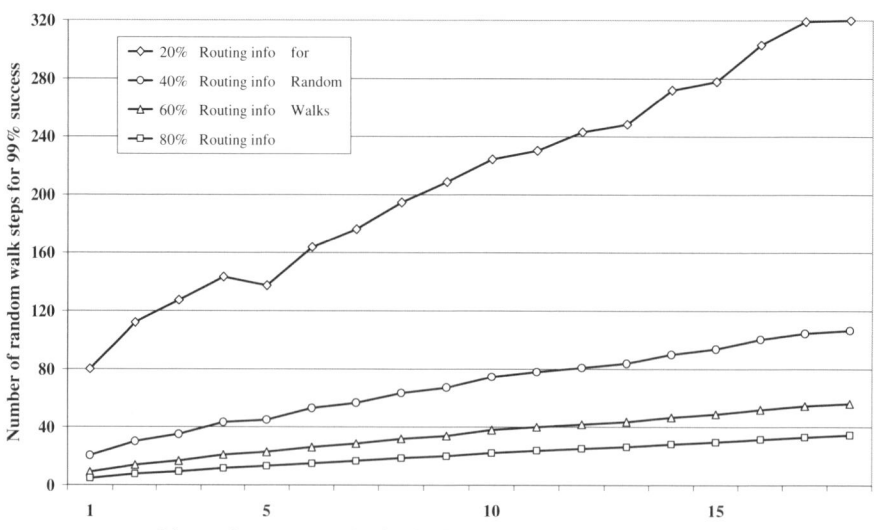

Fig. 7. Biased random walk efficiency: 200 nodes; diameter: 18; mean node degree: 9.5 mean hop distance: 7.0

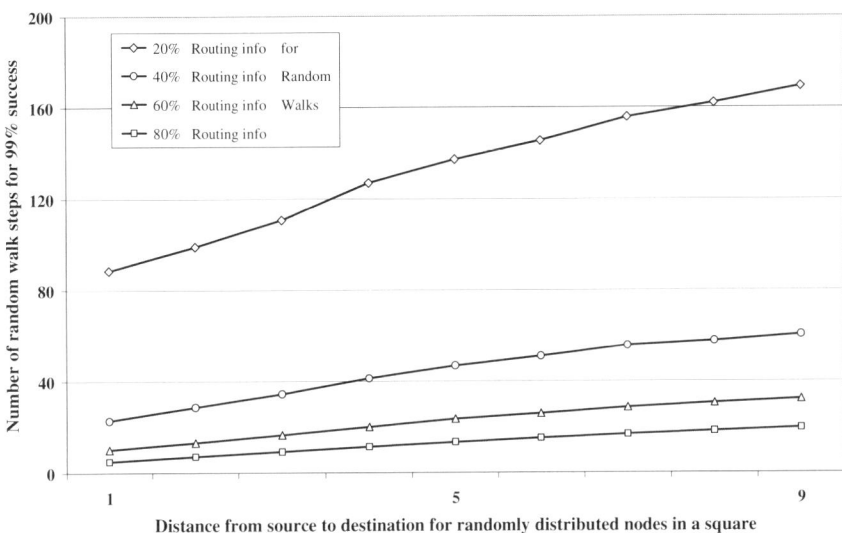

Fig. 8. Biased random walk efficiency: 200 nodes; diameter: 9; mean node degree: 15.6; mean hop distance: 4.2

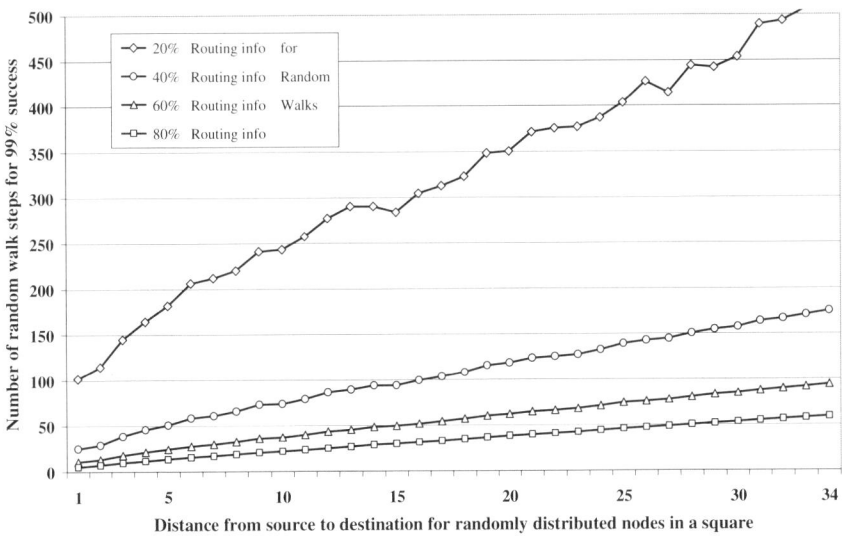

Fig. 9. Biased random walk efficiency: 1000 nodes; diameter: 34; mean node degree: 8.8; mean hop distance: 13.3

The following figures show the results of the random walk efficiency, where curves are included for $\rho = 0.2, \ldots, 0.8$, i.e. for 20%, 40%, 60% and 80% availability of valid routing information at a node. When several neighbors of a node are closer

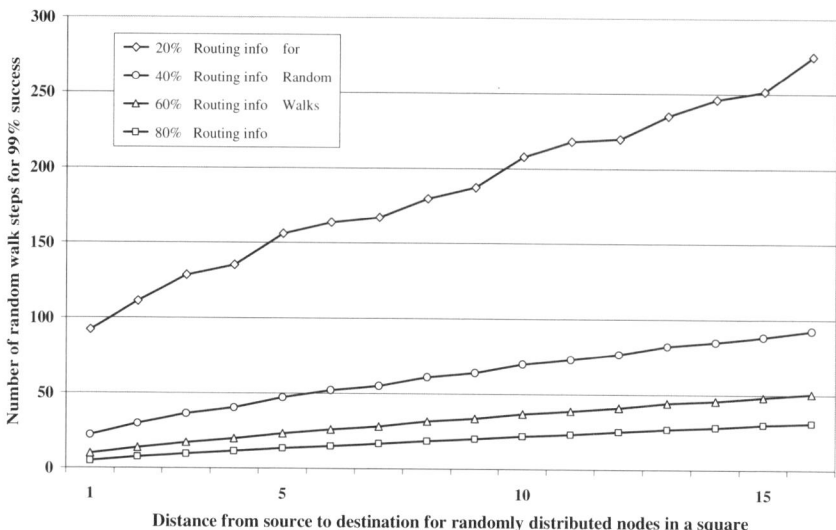

Fig. 10. Biased random walk efficiency: 1000 nodes; diameter: 16; mean node degree: 28.9; mean hop distance: 6.3

to the destination, then one of them is randomly selected as next hop, such that the applied routing scheme is not deterministic even if the next step is not completely taken at random, but directed towards the destination.

In the considered application case of nodes with uniform distribution over a square, the overhead of random forwarding in multihop wireless transmission is significantly below the alternative of flooding, provided that routing information is partially present, enabling to forward messages into the direction of the destination for about 40% or more of the nodes. Thus inserting random steps seems reasonable for routing in unreliable situations in networks subject to frequent changes of the connection topology.

The duration of the corresponding biased random walks naturally improves with the portion of directed steps enabled by valid routing information. Regarding the dependence on the distance from source to destination, a linear increase in number of hops required to reach the destination is visible. When we compare results for the same distance and information level, results are similar in Fig. 7 - Fig. 10, such that further parameters such as the network size, diameter, as well as the mean node and connectivity degree seem to have less influence on the random walk performance.

6   Conclusions and Future Work

Transient analysis can be applied to evaluate the performance of random walks for routing processes in wireless networks in a more fundamental way than using simulations. The work studies biased random walks in planar grid networks and a scenario of uniformly distributed nodes in a square as two basic scenarios for wireless transmission. In addition, a bound on the time to approach the destination is considered, which proves to be fairly tight, especially for grid networks, and is applicable more generally, as shown in related studies of random mobility modeling.

While we investigated basic random walks, several proposals for optimizing random walk performance have not been included. Among them are variants of several random walks in parallel or branching of a walk in a controlled way and the avoidance of a direct step back to the previous node. Target sets consisting of multiple nodes are of interest to model more precise information being available about a target in its surrounding or in cases of queries that can be served by many nodes. Finally, the most efficient way of embedding random steps in routing protocols for low power and lossy networks is open for further study, as will be discussed in the recently established IETF working group for this area.

Acknowledgements

We would like to thank the anonymous reviewers for several valuable suggestions for improvement.

References

1. Avin, C., Krishnamachari, B.: The power of choice in random walks: An empirical study. Computer Networks 52, 44–60 (2008)
2. Beraldi, R.: Service discovery in MANET via biased random walks. In: Proc. Autonomics, Rome, Italy (2007)
3. Broch, J., et al.: A Performance Comparison of Multi-Hop Wireless Ad Hoc Network Routing Protocols. In: Proc. IEEE/ACM Int. Conf. on Mobile Computing and Networking, MOBICOM, pp. 85–97 (1998)
4. Carter, C., Yi, S., Ratanchandani, P.: Manycast: Exploring the space between anycast and multicast in ad hoc networks. In: Proc. MobiCom 2003, San Diego, CA, USA (2003)
5. Czachorski, T., Grochla, K., Pekergin, F.: Diffusion approximation model for the distribution of packet travel time at sensor networks, 4. In: EuroFGI workshop on Wireless and Mobility, Barcelona, Spain, (Janurary 2008),
   http://recerca.ac.upc.edu/eurongi08/ext-abs/8-2.pdf
6. Clausen, T., Jacquet, P.: Optimized Link State Routing Protocol (OLSR), IETF standardization, RFC 3626 (2003), http://www.ietf.org/rfc/rfc3626.txt
7. Coppersmith, D., Feige, U., Shearer, J.: Random walks on regular and irregular graphs. SIAM Journal on Discrete Mathematics 9(2), 301–308 (1996)
8. De Couto, D., Aguayo, D., Chambers, B.A., Morris, R.: Performance of Multihop Wireless Networks: Shortest Path is Not Enough. In: Proc. of the First Workshop on Hot Topics in Networking (HotNets-I), Princeton, New Jersey (2002)
9. Das, S., et al.: Simulation-Based Performance Evaluation of Routing Protocols for Mobile Ad Hoc Networks. Mobile Networks and Applications 5(3), 179–189 (2000)
10. Gkantsidis, C., Mihail, M., Saberi, A.: Hybrid search schemes for unstructured P2P networks. Proc. IEEE Infocom (2005)
11. Gkantsidis, C., Mihail, M., Saberi, A.: Random walks in peer-to-peer networks: Algorithms and evaluation. Performance Evaluation 63, 241–263 (2006)

12. Hasslinger, G.: ISP Platforms under a heavy peer-to-peer workload. In: Steinmetz, R., Wehrle, K. (eds.) Proc. Peer-to-Peer Systems and Applications. LNCS, vol. 3485, pp. 369–381. Springer, Heidelberg (2005)

13. Hasslinger, G., Kempken, S.: Applying random walks in structured and self-organizing networks: Evaluation by transient analysis. PIK journal 31/1; Special issue on self-organizing networks 17-23 (2008)

14. Internet Engineering Task Force: Working Groups on Mobile ad hoc Networks (MANET), http://www.ietf.org/html.charters/manet-charter.html and Routing over Low Power and Lossy Networks (ROLL)
http://www.ietf.org/html.charters/roll-charter.html

15. Johnson, D., Hu, Y., Maltz, D.: The Dynamic Source Routing Protocol (DSR) for Mobile Ad Hoc Networks for IPv4. IETF standardization, RFC 4728 (2007),
http://www.ietf.org/rfc/rfc4728.txt

16. Kunz, T.: On the inadequacy of MANET routing to efficiently use the wireless capacity. In: Proc. IEEE Conf. on Wireless and Mobile Computing, Networking and Communications (WiMob 2005), Montreal, Canada, pp. 109–116 (2005)

17. Kunz, T.: Multicast vs. broadcast in a MANET. In: Nikolaidis, I., Barbeau, M., Kranakis, E. (eds.) ADHOC-NOW 2004. LNCS, vol. 3158, pp. 14–27. Springer, Heidelberg (2004)

18. Kunz, T., Alhalimi, R.: Load-balanced routing in wireless networks: State information accuracy using OLSR. In: Proc. 3rd IEEE Conf. on Wireless and Mobile Computing, Networking and Communications (WiMob 2007), New York, USA (October 2007)

19. Lima, L., Barros, J.: Random Walks on Sensor Networks. In: Proceedings of the 5th International Symposium on Modeling and Optimization in Mobile, Ad hoc, and Wireless Networks (WiOpt 2007), Limassol, Cyprus (April 2007)

20. Lv, Q., Cao, P., Cohen, E., Li, K., Shenker, S.: Search and replication in unstructured peer-to-peer networks. In: Proc. Internat. Conf. on Supercomputing. ACM, New York (2002)

21. Martinez-Arrue, J., Garcia-Escalle, P., Casares-Giner, V.: Location management based on the mobility patterns of mobile users, 4. In: EuroFGI workshop on Wireless and Mobility, Barcelona, Spain (January 2008),
http://recerca.ac.upc.edu/eurongi08/ext-abs/9-2.pdf

22. Qian, L., Kunz, T.: Mobility metrics for adaptive routing. In: Proc. 3rd IEEE ComSoc Conf. on Sensor and Ad Hoc Communications and Networks (SECON), vol. 3, pp. 803–808 (2006)

23. Qin, L., Kunz, T.: On-demand routing in MANETs: The impact of a realistic physical layer model. In: Pierre, S., Barbeau, M., Kranakis, E. (eds.) ADHOC-NOW 2003. LNCS, vol. 2865, pp. 37–48. Springer, Heidelberg (2003)

24. Ripeanu, M., Iamnitchi, A.: Mapping the Gnutella network. IEEE Internet Computing, 50–57 (2002)

25. Terpstra, W., Kangasharju, J., Leng, C., Buchmann, A.: BubbleStorm: Resilient, probabilistic and exhaustive P2P search. In: Proc. ACM SIGCOMM, Kyoto, Japan, pp. 49–60 (2007)

26. Tzevelekas, L., Stavrakakais, I.: Random walks with jumps in wireless sensor networks. In: Proc. of Med-Hoc-Net 2007, Corfu, Greece (June 2007)

27. Villanueva-Pena, P., Kunz, T., Dhakal, P.: Extending network knowledge: Making OLSR a Quality of Service conducive protocol. In: Proc. Int. Conf. on Communications and Mobile Computing, Vancouver, Canada, pp. 103–108 (2006)

28. Zhong, M., Shen, K.: Popularity-biased random walks for peer-to-peer search under the square root principle. In: Proc. 5th Internat. Workshop on P2P systems, IPTPS (2006)

Evaluation of a Cooperative ARQ Protocol for Delay-Tolerant Vehicular Networks*

Julián Morillo-Pozo, Óscar Trullols-Cruces, José M. Barceló-Ordinas,
and Jorge García-Vidal

Computer Architecture Department (DAC),
Techinal University of Catalonia (UPC),
Campus Nord, c/Jordi Girona, 1-3,
08034 Barcelona, Spain
`{jmorillo,trullols,joseb,jorge}@ac.upc.edu`

Abstract. This paper evaluates a Cooperative ARQ protocol to be used in delay-tolerant vehicular networks. The scenario consists in cars downloading information from Access Points along a road. The key difference between proposed Cooperative ARQ protocols is when the cooperation takes place. Simply C-ARQ cooperation occurs in a packet-by-packet basis. In this proposal, that we call DC-ARQ (Delayed Cooperative ARQ), the cooperation is delayed until cars are out of the coverage area of the Access Point. The scheme has been evaluated through simulations. A comparison of DC-ARQ with a baseline case in which no cooperation is used has been performed under different vehicle densities scenarios.

Keywords: Cooperative systems, Disruptive Tolerant Networks, Vehicular Ad-hoc Networks.

1 Introduction

Vehicular Ad-hoc NETworks (VANETs) are a particular case of MANETs in which nodes are vehicles that move following specific patterns (i.e. roads). Important applications of VANETs are: Transportation-related applications and Convenience and Personalized applications including Internet access, hot-spots access, gaming, sharing files or P2P services.

In this paper we focus on delay-tolerant applications, in which cars download information from Access Points (AP) placed on the road.

VANETs are networks characterized by intermittent connectivity and rapid changes in their topology. In the considered scenario, vehicles accessing an AP have few seconds to download information in an environment with high losses. Measurements of UDP and TCP transmissions of vehicles in a highway passing in front of an

---

* This work has been supported by Spanish Ministry of Science and Technology under grant TSI2007-66869-C02-01 and by the NoE EuroFGI of the VI FP of the UE under VNET-3 project.

L. Cerdà-Alabern (Ed.): Wireless and Mobility 2008, LNCS 5122, pp. 157–166, 2008.

AP moving at different speeds report losses on the order of 50–60% depending on the nominal sending rate and vehicle speed; see [1].

In this harsh environment, innovative communication techniques are needed. We believe that cooperative techniques can be beneficial in order to improve the performance of this type of networks and applications. The main objectives of this paper are to test this hypothesis and to compare two possible solutions on this direction.

The main contribution of this work is an evaluation of the Delayed Cooperative ARQ (C-ARQ) scheme to be used in vehicular networks where cars download delay-tolerant information from AP on the road, suffering an intermittent connectivity. In a first general scheme, not addressed to VANET, cooperation among nodes is established in a packet-by-packet basis, [10]. DC-ARQ, on the contrary, cooperation is established in the dark areas, where connectivity with the AP is lost. In this way, AP contact times

To evaluate the scheme we have simulated using NS [2] the case of a straight road with sparse AP giving intermittent connectivity to vehicles. We have also compared the scheme with a baseline case in which no cooperation is used.

The main outcome of our research is that the DC-ARQ protocol can effectively reduce the packet losses of transmissions from access points to cars in a platoon increasing the delivery ratio of the network. This improvement is more spectacular when the vehicle density is high, but also for lower densities DC-ARQ has shown to be an effective technique.

2 Related Work

A performance study in term of losses when vehicles enter the coverage of an access point in a highway and exchange UDP and TCP packets is presented in [1].

Most of the work related to opportunistic vehicular networks deal with opportunistic forwarding strategies, in which nodes schedule the forwarding of packets according to opportunities; see [3], [4], [5]. This scheduling may be based on: historical path likelihoods, [3], packet replication [4], or on the expected packet forwarding delay, [5]. These proposals take as a point of reference epidemic routing [6]. Their main objective is to optimize contact opportunities to forward packets in intermittent scenarios, but they do not consider how to optimize the transference of such information given that you have contacted another node.

Cooperative ARQs are schemes which increase link reliability in data link protocols through the use of node cooperation; see [7], [8], [9], [10]. In [8] authors describe a scheme for improving loss resilience with diversity, focusing on wireless local area networks (WLANs). In [7], authors propose a two-phase communication using a relay node. In [9] authors present a generalization of Hybrid-ARQ where retransmitted packets do not need to come from the original source radio but could instead be sent by relays that overhear the transmission. The job reported in [10] presents a novel frame exchange mechanism between a node and its cooperators for C-ARQ/FC (Cooperative ARQ with Frame Combining).

DC-ARQ, previously proposed in [11], follows the Infostation model, in which nodes transport data and deliver their information during contact times; see [12]. In the case of DC-ARQ hot-spots distributed along roads act as Infostations, while gaps between Infostations are used to interchange packets with other nodes.

Packet-by-packet Cooperative ARQ has as objective to recover immediately lost packets from neighbor vehicles at the cost of bandwidth at contact times. This scheme has sense if different channels are used: one to download information from AP and other to cooperate. Delayed Cooperative ARQ (DC-ARQ) has as objective of not wasting bandwidth at contact times and waits to dark areas in which vehicles have not connectivity with AP to cooperate and recover as much packets as possible from neighbor nodes. This scheme has been proposed and tested in a simple prototype with 3 vehicles in an urban scenario, [11]. The tests were performed using ping packets sent from an AP to each of the 3 vehicles and then, these vehicles recovered losses after leaving AP coverage. The work showed that vehicles were able to recover in some experiments around 50% of the packets lost. However, we did not test the prototype in denser scenarios due to the difficulty in scaling experiments. Furthermore, cooperation scenarios were not compared not cooperation scenarios.

3   Cooperative ARQ for Delay-Tolerant Vehicular Networks

Consider Figure 1 in which vehicles want to download information from the Internet through AP distributed along a road. Due to the harsh conditions produced in VANET, the losses produced in such environment are high. Reference [1] reports experiments on a highway in which vehicles passing in front of an AP moving at different speeds have losses on the order of 50-60% depending on the nominal sending rate and vehicle speed.

AP acts as custodians in a Delay/Disruptive Tolerant Network (DTN), [13], keeping files for nodes. It is not the purpose of this paper to define the DTN architecture. We will assume that there is a server that schedules which files are stored at each AP and which and how packets are sent to the vehicles.

Although and end-to-end ARQ mechanism is needed to recover packets, we may use the broadcast nature of the media to improve packet recovery. Cooperative techniques may improve performance in terms of packet losses. When a vehicle receives a packet, its neighbor nodes may independently receive the packet. In this way, some of the vehicles in the platoon traveling with the destination node may have some of the lost packets. Nodes, then, may cooperate to recover packet losses minimizing the number of end-to-end retransmissions.

3.1   Delayed Cooperative ARQ (DC-ARQ)

The key idea of this scheme is to delay retransmissions coming from cooperators until the platoon is on a dark area out of coverage of any AP. This scheme operates into three phases, see Fig. 1:

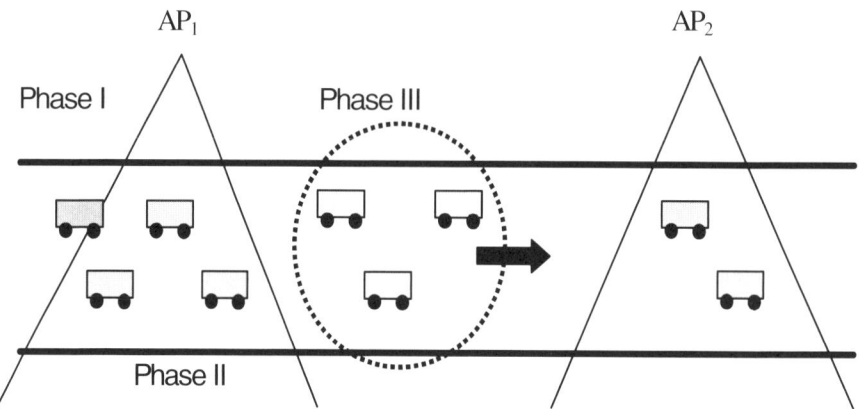

Fig. 1. VANET scenario

3.1.1 Association Phase

Vehicles detect AP along the road and register its willingness to download packets. Suitable mechanisms for detecting in-range AP, association and authentication of vehicles reaching a given AP must be provided. Those mechanisms can have a major importance on the overall performance, but are not specific of the use or not of cooperation, and thus we leave them out of the scope of this paper. It can be assumed, for example, that vehicles are equipped with WAVE (Wireless Access in Vehicular Environments) IEEE 802.11p cards. WAVE architecture provides mechanisms to access WAVE Base Stations (AP) in vehicular networks. In our framework, AP periodically broadcast hello beacons each 0.2 seconds, identifying their zone. Each node that receives the beacon knows it is in coverage and will reply with a registration packet containing its identification.

3.1.2 Reception Phase

We first describe the basic Packet-by-Packet Cooperative ARQ mechanism and then modified it to consider Delayed Cooperative ARQ. Vehicular nodes receive data from the AP. Each car receives its data but also buffers the packets addressed to other cars in the platoon that consider it as cooperator. The cooperation relationship is established through the exchange of HELLO messages broadcasted periodically by the vehicular nodes. The first function of a HELLO message sent by a node x is to allow other nodes to know about the presence of x. Other vehicular node y in the platoon will add x to its list of cooperators (if x is not already a cooperator of y) when receiving this HELLO message. The second function of a HELLO message sent by a node x is to notify other nodes about the fact that they have to act as cooperators of x. For this second function, each HELLO message contains the list of cooperators of the sending node. In our example, the next HELLO message sent by y will contain x in the list of cooperators. In this way, x will be aware of the fact that y considers it as cooperator and will act accordingly (buffering packets addressed to y).

The list of cooperators contained in the HELLO messages also indicates the order in which cooperators should act when the destination node fails to receive the packet from the AP: this is to avoid collisions; when cooperators detect that the destination node has not received correctly a packet after the ACK timeout, each cooperator will wait a fixed back-off depending on this assigned order, before retransmitting the packet. A retransmission from the AP will be only needed in case that any cooperator can not deliver the correct copy of the packet to the destination node. Note that we do not focus on the cooperators selection algorithm, so this is left out of the scope of this document.

For the case of DC-ARQ, cooperators do not retransmit the packets while they are in this phase. Retransmission (i.e. cooperation) of packets will be delayed until the platoon of cars is out of the coverage area in a new phase that we call *Cooperative-ARQ phase* (see next section 3.2.3).

In the considered scenario, data flow is always from the AP to the vehicular nodes, and no retransmissions are used. We avoid retransmissions at the hope that other cars in the platoon (i.e. cooperators) will receive packets incorrectly received by the destination and will help it in the Cooperative-ARQ phase, without the need of wasting the useful time in coverage with the AP in retransmissions. In this way the channel can be used by the AP to transmit as much new data addressed to the cars as possible, thus reducing the downloading time and increasing the effective data rate.

3.1.3 Cooperative-ARQ Phase

When the cars leave the AP range, they enter into the Cooperative-ARQ phase. After some time (e.g. equivalent to 10 beacons) without receiving beacons from the AP, a node considers that is out of the AP coverage. At this point, every node checks which packets it has failed to receive correctly from the AP and starts to request them to other vehicular nodes (i.e. to its cooperators), in an attempt to recover all packets from the first to the last received from the AP. The process is the following: (i) A node x broadcasts a REQUEST packet for each started block that it has failed to complete from the AP with its packet's received bitmap. (ii) When receiving this REQUEST, each cooperator of x will check if it has any packet from the requested block buffered (it has received the packet correctly from the AP in the previous phase). (iii) If it has some packets, it will send the packets to x (unless other cooperator sends it before).

This process will be repeated while the node receives any packet from its requests, and ends after the tenth time that any cooperator haven't replied or when it enters in range of a new AP, meaning that it comes into reception mode (Reception phase of the protocol operation), and the whole cycle starts again.

Note that the end-to-end block ARQ still is working on top of the cooperative ARQ mechanisms. The scheme trade-offs among delaying the recovery of packets from cooperators until vehicles reach dark areas in which there are no AP contacts and recovering the packets from the higher layer ARQ protocol.

4 Performance of Delayed Cooperative ARQ

The simulations were performed with the standard version of the ns-2.31 simulator [2]. Each AP generates packets to vehicles in its coverage area in round robin basis. Vehicles request files of 10 MB and packets have a size of 1 KB.

Vehicles travel 30 Km on a two-way highway with two lanes per direction. The road network infrastructure consists of 5 Access Points placed every 6 Km. Vehicles move with constant speed, randomly chosen from a uniform distribution between 70–90 km/h on the right lane and between 90-120 km/h on the left lane. Vehicle density is modeled with exponential distribution of parameter λ_1 vehicles/s in the right lane and λ_2 vehicles/s in the left lane. These rates consider the maximum number of vehicles following vial rules (e.g. security distances of 80 meters traveling at 90 Km/h, 100 meters traveling at 100 Km/h, etc). Using these consideration, $\lambda_1=0.25$ vehicles/s in the right lane and $\lambda_2=0.2$ vehicles/s in the left lane would fill the highway with the maximum number of vehicles traveling at 100 Km/h and at 120 Km/h at each lane. Higher densities may be achieved lowering vehicle speeds (then the security distance between vehicles is lower). In the graphs we normalize λ_i (i=1,2) and for clarity we define parameter α as vehicle density, being $\alpha=1$ the higher density corresponding to vehicles with 120 Km/h as maximum speed at line 1 and 90 Km/h as maximum speed at line 2. Lower values of α indicate a decrease in vehicle density.

For the physical layer we consider IEEE802.11a at a rate of 3 Mb/s. Nodes use omni-directional antennas, see Table 1.

Table 1. Configuration parameters

Antenna	Omnidirectional
Frequency	5.9 GHz
RxTh	-95 dBm
CSTh	-96 dBm
Antenna Gain	2.512 dB
TxPower	9.95 dBm

We have used Nakagami as propagation model. This model is used to predict signal attenuation in fading environments and has already been used in vehicular scenarios, [14]. The Nakagami probability density function, [15], defines a distribution of the power x of the received signal:

$$f(x;m,\Omega) = \frac{m^m}{\Gamma(m)\Omega^m} x^{m-1} e^{-\frac{m}{\Omega}x}$$

Where, Γ denotes the Gamma Function, Ω denotes the average received power, m is the Nakagami parameter and both Ω and m depends on the distance between transmitter and receiver. High values of the m-parameter (with m>1) introduces a variability on the average power reception similar to two ray-model while a value of m=1 introduces Raleigh distribution variability. Lower values of the parameter m worsen channel performance. That allows to define a reception similar to two-ray mode in short distances with reception powers that depends on the d^{-2} and reception similar to fading models in larger distances with reception powers that depends on $d^{-\gamma}$ ($\gamma >2$).

Figure 2 shows delivery ratio for of two path loss scenarios and two antenna gains. The scenarios considered are specified in Table 2. Scenario A considers a two-ray model in short distances (d<200 meters). At distances higher than 200 meters, the reception power decreases at a power of $\gamma >3.8$. The fading channel is the same at all distances: a high loss fading channel of m=0.5. Scenario B is very similar to scenario A, but with different fading channel depending on the distance. Distances lower than 80 meters have a Raleigh fading channel, from 80 to 200 meters the channel is worst (m=0.75) and finally at higher distances (d>200 meters) m=0.5 (an even worst fading channel). As a summary Scenario A is more pessimistic than Scenario B.

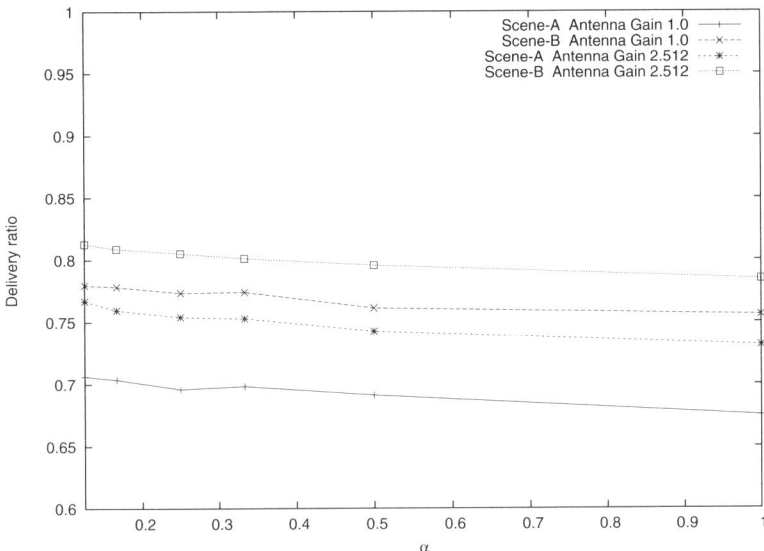

Fig. 2. Delivery ratio as a function of car density

Table 2. Fading Scenarios

		d < 80	80 < d < 200	200 < d
Scenario A	**m**	0.5	0.5	0.5
		d < 200	200 < d < 500	500 < d
	γ	1.9	3.8	3.8
Scenario B	**m**	d < 80	80 < d < 200	200 < d
		1.0	0.75	0.5
		d < 200	200 < d < 500	500 < d
	γ	1.9	3.8	3.8

Remember that the α depicted on the x-axis means the normalized car density with respect to the maximum possible. As expected, results show a worst delivery ratio on scenario A and a straight line in all the scenarios indicating independence with respect the vehicle density. From Fig. 2 it can also be observed the impact of the antenna gain in the delivery ratio.

Figure 3 shows the packet delivery ratio as a function of vehicle density for the base-line case in which no cooperation is used and for the case in which the proposed cooperation scheme (DC-ARQ) is utilized. The two curves correspond to Scenario B with an antenna gain of 1dBm (these parameters are the defaults in ns-2 for the Naka-gami channel). The first curve (i.e. no cooperation) shows almost a straight line indi-cating that packet losses are independent of vehicle density, as seen on previous Figure 2. That means that as long as we are not using other cars to improve the recep-tion of the receivers, they do not affect to the overall delivery ratio. Note the difference with the case of using DC-ARQ: in this case, as long as the number of cars on the road

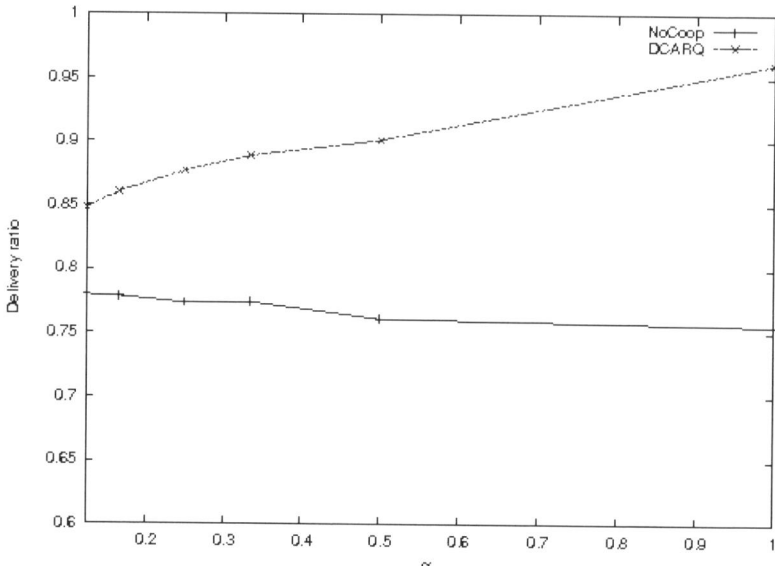

Fig. 3. Delivery ratio as a function of car density (Scenario B)

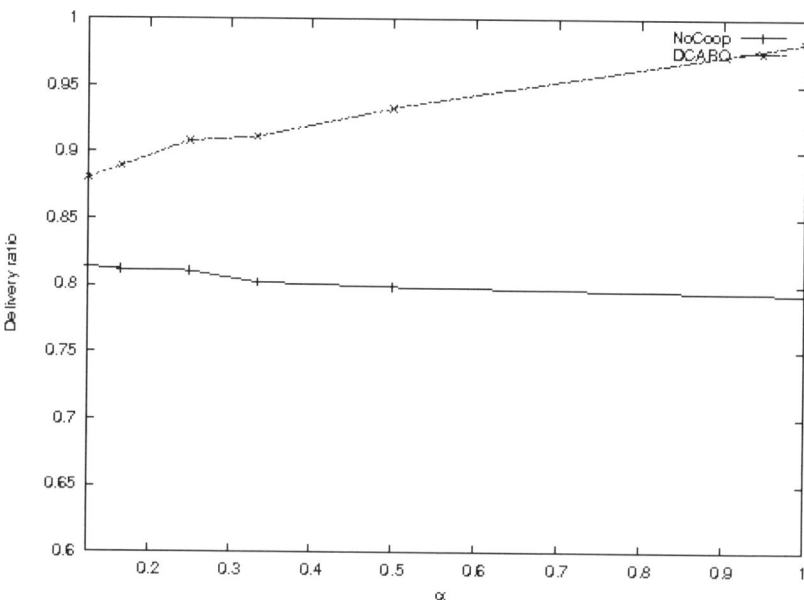

Fig. 4. Delivery ratio as a function of car density (m=1)

(i.e. vehicle density) increases, receiving nodes can benefit from cooperation. Note that at maximum node density DC-ARQ allows for an improvement of the delivery ratio from 0.76, in which no cooperation is used, to 0.96. Also interesting, is the improvement that DC-ARQ allows even for lower densities.

Figure 4 depicts a slightly better scenario in which the parameter m is set to 1 for all the distances (i.e. cars are always facing a Rayleigh channel). We can observe the same behavior than that in Figure 3. The only difference is that as long as the channel conditions are better, both curves (i.e. no cooperation and DC-ARQ) perform better). It is interesting to see, however, how the delivery ratio of DC-ARQ approaches 1 in this case for a full node density. More scenarios have been simulated, and in all cases the same behavior has been obtained.

From Figures 3 and 4 we may extract some interesting conclusions: (i) DC-ARQ allows the recovery of lost packets from neighboring vehicles. This recovery may be performed in dark areas in which there is not Internet access. In this way, use of bandwidth during contact times is optimized. (ii) DC-ARQ recovers packets but is not an end-to-end ARQ mechanism. That means that is not able to guarantee the whole reception of a file. (iii) DC-ARQ is able to recover with high probability almost all packets when vehicle density is higher. The idea is that in high vehicle density scenarios there always are vehicles in the AP communication range. These vehicles will receive packets destined to other vehicles with high probability and afterwards cooperate with them to deliver them. This can be seen in Figures 3 and 4 where in high scenarios the recovery of packets is quite high. However, in low vehicle density scenarios, there are few or none vehicles in the AP communication range that may receive packets destined to other vehicles. In this case, waiting to the end of the communications range to recover packets wastes the possibility to recover packets using an end-to-end ARQ during AP contact times. In any case, an end-to-end ARQ mechanism is needed to recover those lost packets, even in high vehicle density scenarios.

From these conclusions, we think in three possible solutions to analyze in further studies: (a) an adaptive scheme in which AP switch from end-to-end ARQ to DC-ARQ when vehicle densities are high on its coverage. The end-to-end ARQ would be enabled when vehicles enter next AP coverage. (b) using different channels in which reception of information and cooperation is established. (c) Network coding would allow synchronization in the reception of packets and therefore would add robustness.

5   Conclusions

This paper has evaluated Delayed Cooperative ARQ (DC-ARQ) mechanism in a Vehicular Ad Hoc Network (VANET). The paper analyzes a Delay/Disruptive Tolerant Network scenario in which vehicles download information from access points distributed along roads. Optimizing the use of bandwidth during contact times is a key factor in this kind of networks. DC-ARQ waits until vehicles leave AP coverage to cooperate in dark areas not wasting AP bandwidth resources in recovering packets. Simulation shows that DC-ARQ may recover high amount of packets in high vehicle density scenarios. However, in low vehicle density scenarios cooperation is not able to recover too many packets. This is due to the lack of cooperators. In these cases, contact times should be used to download as much as possible information together with an end to end recovery mechanism.

References

1. Ott, J., Kutscher, D.: Drive-thru Internet: IEEE 802.11b for 'Automobile' users. In: IEEE INFOCOM (2004)
2. Network Simulator ns-2, http://www.isi.edu/nsnam/ns/
3. Burguess, J., Gallagher, B., Jensen, D., Levine, B.N.: MaxProp: Routing for Vehicle-based Disruption Tolerant Networks. In: 25th Conference on Computer Communications (IEEE INFOCOM), Barcelona, Spain (April 2006)
4. Balasubramanian, A., Levine, B.N., Venkataramani, A.: DTN Routing as a Resource Allocation Problem. In: ACM SIGCOM 2007, Kyoto, Japan (August 2007)
5. Zhao, J., Cao, G.: VADD: Vehicle-Assisted Data Delivery in Vehicular Ad Hoc Networks. In: 25th Conference on Computer Communications (IEEE INFOCOM), Barcelona, Spain (April 2006)
6. Vahdat, Becker, D.: Epidemic Routing for Partially Connected Ad Hoc Networks, Technical Report CS-200006, Duke University (April 2000)
7. Zhao, B., Valenti, M.C.: Practical Relay Networks: A Generalization of Hybrid-ARQ. IEEE Journal on Selected Areas in Communications 23(1) (January 2005)
8. Miu, A., Balakrishnan, H., Koksal, C.E.: Improving Loss Resilience with Multi-Radio Diversity in Wireless Networks. In: ACM Mobicom 2005(September 2005)
9. Dianati, M., Ling, X., Naik, K., Shen, X.: A Node Cooperative ARQ Scheme for Wireless Ad-hoc Networks. In: IEEE Wireless Communications and Networking Conference (2005)
10. Morillo, J., Garcia-Vidal, J.: A Low Coordination Overhead C-ARQ Protocol with Frame Combining. In: 18th Annual IEEE International Symposium on Personal, Indoor and Mobile Radio Communications (IEEE PIMRC 2007), Athens, Greece (September 2007)
11. Morillo-Pozo, J., Trullols, O., Barceló, J.M., García-Vidal, J.: A Cooperative ARQ for Delay-Tolerant Vehicular Networks (to be published in DTMN 2008)
12. Small, T., Haas, Z.J.: The Shared Wireless Infostation Model – A new Ad Hoc Networking Paradigm (or Where there is a Whale, there is a Way). In: MobiHoc 2003, Annapolis, Maryland, USA (June 2003)
13. Fall, K.: A Delay-Tolerant Network Architecture for Challenge Internets, IRB-TR-03-2003, Intel Research Berkeley (Feburary 2003)
14. Torrent-Moreno, M.: Inter-Vehicle Communications: Achieving Safety in a Distributed Wireless Environment, Dissertation. Universitätsverlag Karlsruhe (2007)
15. Zhang, W., Moayeri, N.: Classification of Statistical Channel Models for Local Multipoint Distribution Service using Antenna height and directivity, IEEE 802.16 working group contribution IEEE802.16.1pc00/07 (January 2000)

A Seamless Vertical Handover Approach*

Rastin Pries, Dirk Staehle, Phuoc Tran-Gia, and Thorsten Gutbrod

University of Würzburg, Institute of Computer Science, Germany
{pries,staehle,trangia}@informatik.uni-wuerzburg.de

Abstract. Today, wireless network devices are equipped with multiple access technologies like UMTS and Wireless LAN. The handover between the technologies has been widely studied in literature. Several of these research papers build their handover mechanisms on top of existing protocols like Mobile IP. These protocols operate on the network layer and recognize link layer changes only by timeouts. Consequently, if no cross layer approach is implemented, a seamless vertical handover between Wireless LAN and UMTS cannot be provided.

In this paper, we introduce a vertical handover protocol based on a tightly-coupled network architecture. With this architecture, it is possible to perform the handover on the link layer and thus, reducing the handover delays. The protocol is implemented in a simulation environment and the results reveal that a seamless vertical handover can be performed with blackout times of only a few milliseconds.

Keywords: vertical handover, UMTS, Wireless LAN, tight coupling.

1 Introduction

The immense growth of wireless access technologies and the increased demand for mobile Internet access leads from the currently existing third generation cellular networks to the next generation mobile networks. These mobile networks will most likely be based on a packet-switched architecture with a diversity of access technologies. With such an architecture, the 3G mobile networks can easily be extended by other IP based wireless access technologies like *Wireless Local Area Network* (Wireless LAN) or *Worldwide interoperability for Microwave Access* (WiMAX). The design and purpose of these various wireless networks are completely different. However, exactly their diversity of characteristics complement one another which makes the integration attractive. On the one hand, we already have the ongoing *Universal Mobile Telecommunications System* (UMTS) as a representative of 3G cellular networks which provides wide-area coverage with high mobility and a comparatively low bandwidth. On the other hand, several technologies like Wireless LAN provide a lower coverage area but offer a high bandwidth. Combining these technologies creates a ubiquitous wireless network with local hotspots to support the user with high speed services.

Several approaches have been published [1,2,3] showing how to combine Wireless LAN and UMTS and how to perform a handover between these technologies. Almost

---

* This work is funded by Deutsche Forschungsgesellschaft (DFG) under grant TR 257/19-2. The authors alone are responsible for the content of the paper.

L. Cerdà-Alabern (Ed.): Wireless and Mobility 2008, LNCS 5122, pp. 167–184, 2008.

every approach is based on Mobile IP [4] and its extensions [5,6]. However, simulations have shown [7] that the handover performance with Mobile IPv4 is really low even when performing a handover within one technology. In [8], Mobile IPv6 [9] is used to perform a vertical handover between Wireless LAN and UMTS. The results show handover delays between 2 and 4 seconds which is not sufficient if real-time applications have to be supported. The low performance results from the operation on the network layer. Mobile IP sends out messages periodically to look for connection changes. Only if three consecutive messages are lost, a handover will be performed. The consequences are handover delays of several seconds. A possibility to reduce the handover delays is a cross layer approach like shown in [7,10]. Here, every link layer change is reported to the network layer, who can immediately react to this change and thus, the blackout times are reduced to less than one second.

Another possibility to perform a vertical handover is by using the *Session Initiation Protocol* (SIP) presented e.g. in [2]. However, similar to the Mobile IP approaches, a loose coupling architecture is used, meaning that the Wireless LAN is not integrated into the UMTS network. This results in handover delays of 1 to 4 seconds when switching from Wireless LAN to UMTS. The SIP is also used for the signaling of the *IP Multimedia Subsystem* (IMS) [11]. Munasinghe et al. [12] show how to perform a vertical handover between Wireless LAN and UMTS using IMS. However, no results are presented.

Individual vertical handover protocols are proposed in [13] and [14]. The first approach by Shenoy and Montalvo is rather complex but the handover delay is with less than 300 ms a lot faster compared to the other approaches. The second approach uses the *Stream Control Transmission Protocol* (SCTP) for the handover signaling. The handover delay is with around 220 ms the lowest of all presented solutions. However, it is assumed that the complete association on the lower layers has already been performed and the 220 ms are the blackout time with no connection at all.

Our proposed vertical handover protocol lays the foundation for the policy-based vertical handover suggested in [15]. The prerequisites for the policy decision metric is a fast and reliable vertical handover which supports even *Quality of Service* (QoS) demanding applications like voice over IP. In order to meet these requirements, the two networks need to have the same authentication environment to reduce the number of signaling messages, which we realized with a tightly coupled network architecture. In the approach, the *Serving GPRS Support Node* (SGSN) receives and evaluates all network relevant parameters and initiates the handover based on a set of policies. In this paper, we will show the performance of the vertical handover between a UMTS network and a Wireless LAN based on the IEEE 802.11g [16] standard and the results show that the protocol is, with only a few milliseconds blackout time, fast enough to support real-time applications.

The rest of this paper is organized as follows. The second section describes the vertical handover network architecture by showing all possibilities of how to combine the networks. Furthermore, the new protocol stack of the *Mobile Equipment* (ME) is shown. Section 3 introduces the vertical handover protocol. In Section 4, the performance of the protocol is presented and validated through simulation. Finally, this paper is concluded in Section 5.

2   Network Architecture

There are three possibilities of how to integrate Wireless LAN and UMTS: loose coupling, tight coupling, and very tight coupling. This describes how Wireless LAN is integrated into the UMTS network. The current approach of mobile network operators to offer the networks separately without coupling and without the possibility of using a vertical handover will not be regarded as an integration.

2.1   Loose Coupling

Loose coupling means that the Wireless LAN *Access Point* (AP) is connected to the *Gateway GPRS Support Node* (GGSN) or the Internet as shown in Fig. 1. The problem with this architecture is that switching the access networks means changing the IP address because the IP packets have to be sent to the new location. However, changing the address means that the existing session will be disrupted. Therefore, a vertical handover approach like Mobile IP is needed. Furthermore, authentication, accounting, and mobility management mechanism for the Wireless LAN have to be realized. Thus, loose coupling is a good choice if you want to use your private Wireless LAN because mechanism like authentication or accounting are done separately at the wired Internet link. Although, if absolutely necessary in case of a Wireless LAN Internet service provider, an overlay network like the IEEE 802.21 [17] can handle this. The huge advantage of loose coupling is that the UMTS architecture can remain unaffected and in the far future it will be easier for the network service provider to exchange the UMTS network with a new technology because only one interface for the new network has to be integrated.

Because of handling the vertical handover on the network layer with solutions based on Mobile IP or a similar mechanism, a vertical handover in a loose coupling architecture is slower than in a tight coupling architecture.

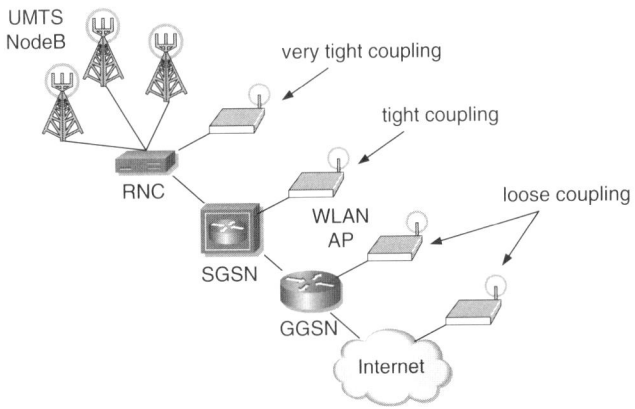

Fig. 1. Different network coupling architectures

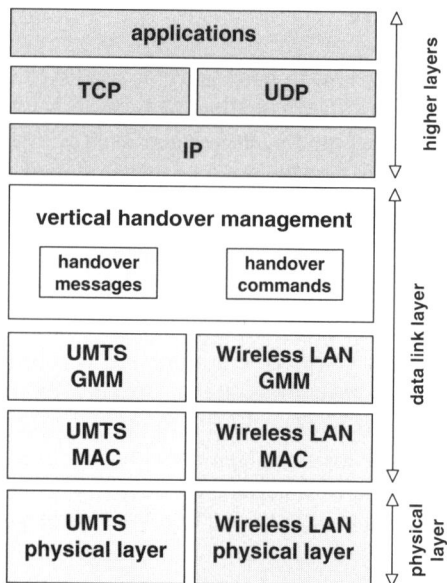

Fig. 2. Protocol stack of the Mobile Equipment

2.2   Tight Coupling

When the Wireless LAN Access Point is connected directly to the *Serving GPRS Support Node* (SGSN) this is called tight coupling. The Wireless LAN is integrated into the UMTS network and all UMTS features like authentication, billing, and mobility management are done by the UMTS network. This way, the SGSN must be able to detect and handle directly connected Wireless LAN Access Points which means that additional changes have to be performed. The Wireless LAN Access Point has to support the functionality of a *Radio Network Controller* (RNC), for example it has to establish IP-tunnels to the SGSN to deliver the packets coming over the Wireless LAN.

2.3   Very Tight Coupling

As in the tight coupling architecture, the Wireless LAN Access Point is integrated into the UMTS network. The difference between the two architectures is the point of integration. In a very tight coupling architecture the Wireless LAN Access Point is connected to the RNC and has to provide the functionality of a NodeB, while in the above discussed tight coupling architecture, the Wireless LAN Access Point is connected to the SGSN and has to provide the functionality of an RNC. The disadvantage of this architecture is that incoming IP packets from the ME have to be encapsulated in smaller packet data units to be transported to the RNC. This can be avoided by the tight coupling architecture.

Loose Coupling as well as tight or very tight coupling requires extensions of the ME regarding sending or receiving network information and executing the handover.

2.4   Mobile Equipment Protocol Stack

In order to perform a seamless vertical handover, the Mobile Equipment has not only to be equipped with two interfaces, but further with the possibility to switch between the interfaces. In a UMTS network, the communication between an ME and an SGSN is internally addressed by a temporary mobile subscriber identity for services provided through the SGSN (P-TMSI) as standardized by 3GPP in [18]. As we are using a tightly coupled approach, the Wireless LAN interface has to be equipped with a P-TMSI as well. Therefore, the Wireless LAN interface of the Mobile Equipment is extended with a GPRS Mobility Management (GMM) module which takes care of attaching the Mobile Equipment to the GPRS network with Wireless LAN and it handles the *Packet Data Protocol* (PDP) control and further functionalities for using GPRS. The complete protocol stack of the Mobile Equipment is shown in Fig. 2. Besides the Wireless LAN GMM extension, a vertical handover management module is placed above the UMTS and Wireless LAN GMM modules. This module is responsible for executing the vertical handover messages and commands and to forward upper layer messages to the currently active interface.

3   Vertical Handover Protocol

We decided to use the tightly coupled approach for the following reasons. First, the very tightly coupled approach requires packet fragmentation at the Access Point in order to transmit them to the RNC. Second, a loosely coupled approach requires extra signaling messages for authentication. Besides architecture and entities, the vertical handover stands and falls with its protocol. On the one hand, it is important to minimize the signaling to a required level in order to ensure a fast handover. This is particularly relevant, when the vertical handover has to be performed in a highly loaded network, for example to transfer a few users to another network which is not operating at full capacity. On the other hand, a minimum signaling traffic between the core network and the Mobile Equipment may result in incomplete handovers due to packet loss or interruption of sessions.

Additionally, in the special case of Wireless LAN and UMTS, status information has to be transmitted between the Mobile Equipment and the core network to be able to decide when a vertical handover should be initiated. This relies on the fact that Wireless LAN provides the information in the Mobile Equipment and UMTS stores this information in the access and core network, namely in the Radio Network Controller and in the Serving GPRS Support Node. Due to the fact that our vertical handover is network initiated, controlled by the SGSN, additional Wireless LAN status information is sent from the ME to the SGSN. The handover decision module in the SGSN requests as many updates of the location, signal strength, available networks, and similar information as possible to judge the current situation. However, for the network it is important to keep this signaling as low as possible.

Altogether, a trade-off has to be found between these aspects to preserve stability for the vertical handover process itself and simplicity at the same time. It has to be ensured that the vertical handover can be performed even when packet loss or long delays occur or furthermore, the vertical handover can be aborted when the destination network becomes unavailable. In order to ensure a reliable vertical handover, several timers are initiated during the process. These timers are used for fallback solutions if anything unexpected happens during the handover process. The timers and the handover messages itself are not part of the 3GPP or the IEEE 802.11g standard and have been developed by ourselves. However the connection establishment and the connection release messages are implemented as described in the standards.

3.1   Handover from Wireless LAN to UMTS

The complete vertical handover process from the Wireless LAN, based on the IEEE 802.11g standard, to the UMTS network is shown in Fig. 15 in the Appendix A. At the top the involved entities are presented. To keep it simple, only the important steps are shown and minor aspects of the vertical handover, like intercommunication and tunnel management between the SGSN and GGSN, are suppressed in this figure although they are implemented in the simulation.

At the left, there is the ME which is able to connect via the air interface to the Wireless LAN Access Point or NodeB which together with the RNC forms the *UMTS Terrestrial Radio Access Network* (UTRAN). The Wireless LAN Access Point and the RNC are connected to the SGSN. All traffic sent over the Wireless LAN is marked gray, traffic sent over the UMTS network is black and traffic within the access or core network and other traffic is illustrated with dashed arrows. At the left margin, the timers used by the ME are drawn and at the right margin, there are the timers of the SGSN. Finally, tunnel operations are marked using dotted arrows.

We can clearly recognize the three different parts of the vertical handover process in Fig. 15: first, the UMTS network connection is established, then the network switching is done, and at the end the Wireless LAN connection is released. So let us take a closer look at the establishment of the Wireless LAN connection which is for better reading separately illustrated in Fig. 3.

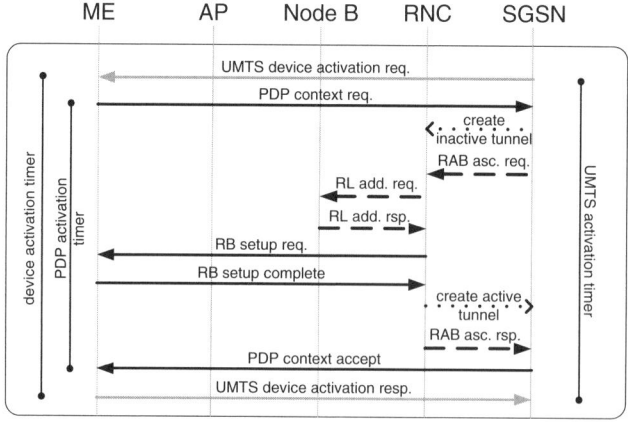

Fig. 3. UMTS connection establishment

UMTS connection establishment

Due to the fact that a vertical handover from Wireless LAN to UMTS should be performed, an existing Wireless LAN connection can be expected. Using this connection, a UMTS device activation request is sent from the SGSN to the ME as shown in Fig. 3. The ME now looks what kind of *Packet Data Protocol* (PDP) context is already active over Wireless LAN. For every quality of service class, a PDP context request is transmitted to the SGSN via the UMTS shared channel in order to establish PDP contexts for these quality of service classes.

If the SGSN receives one of these requests, it creates an inactive data tunnel to the RNC and sends a radio access bearer association request to the RNC. A *Radio Access Bearer* (RAB) is a service which is provided to the non-access stratum by the access stratum for transfer of user data between the ME and the core network. For this service, a connection between the RNC and the NodeB is necessary and additionally a connection between the ME and the RNC. These are the next steps to complete the RAB association.

The RNC transmits a radio link addition request to the NodeB which is a request to add a new radio link to the active set, which is a set of radio links simultaneously involved in a specific communication service between an ME and a NodeB. When this radio link addition request is successfully answered by a radio link addition response, a connection between the RNC and the ME is needed, so the RNC sends a *Radio Bearer* (RB) setup request to the ME. The radio bearer is the service provided by the layer 2 for transfer of user data between the Mobile Equipment and the UTRAN. When the radio bearer setup is successful, an RB setup complete is sent from the ME to the RNC which then creates an active data tunnel to the SGSN. After that, the RNC confirms the established RAB with a radio access bearer association response. The PDP context for one quality of service class is now active and the SGSN confirms this to the ME QoS class with a PDP context accept message.

When every quality of service class whose PDP context has to be established is active, the Mobile Equipment is connected to both networks and transmits a positive UMTS device activation response to the SGSN. All communication of the UMTS connection establishment has to be done over UMTS except the device activation request and response message. UMTS uses four quality of service classes, so the PDP context request is simultaneously transmitted up to four times. Without loss of generality and in order to avoid an overcrowded diagram, only one PDP context is established in our figures.

Vertical handover - network switching from Wireless LAN to UMTS

Fig. 4 shows the second section of the vertical handover. After the successful connection establishment, the SGSN will initiate the vertical handover in the narrower sense, thus the switching of the networks. Therefore, a handover initiation message is transmitted from the SGSN to the ME. The receiving causes the ME to switch the sending device. It transmits further packets with the UMTS device instead of the former used Wireless LAN interface.

To indicate the switching of the devices to the SGSN, the Mobile Equipment sends a handover response message to the SGSN. When the Serving GPRS Support Node processes this information, it activates the previously created but inactive IP-over-IP

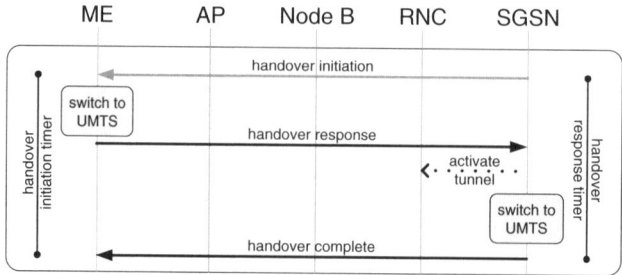

Fig. 4. Network switching from Wireless LAN to UMTS

data tunnel from the RNC to itself. Now, the SGSN is able to apply this data connection to the ME via the RNC in order to switch the networks. The Serving GPRS Support Node has to assign one internal used P-TMSI to every public IP address, which is used for addressing the device from external networks. Hence, the Serving GPRS Support Node updates the entry of the Mobile Equipment's IP address in the SGSN address mapping table with the P-TMSI of the Mobile Equipment's connected UMTS device. From that moment, all incoming traffic for the ME is routed over the UMTS network and a handover complete message is transmitted to the ME. The network switching has been completed and it is necessary to release the Wireless LAN connection which is not needed anymore.

Wireless LAN connection release

When the network has switched to UMTS, it is advisable to release the Wireless LAN link, primarily considering power saving issues. This process is displayed in Fig. 5. Accordingly, the Serving GPRS Support Node removes the data tunnel to the Wireless LAN Access Point and sends a Wireless LAN device deactivation request to the Wireless LAN Access Point. After processing it, such a request is also transmitted to the ME from the Wireless LAN Access Point, which then removes the tunnel to the Serving GPRS Support Node and the ME from its *Access Control List* (ACL). The arrival of the Wireless LAN device deactivation request over the Wireless LAN interface tells the ME to deactivate the connection to the Wireless LAN Access Point.

To deactivate all connections to the Wireless LAN Access Point, the ME sends a Wireless LAN disassociation message to the Wireless LAN Access Point which on its

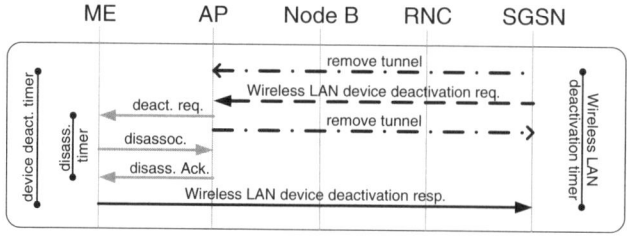

Fig. 5. Wireless LAN connection release

part answers with a Wireless LAN disassociation acknowledgment. At this point, the Wireless LAN connection is released and the ME informs the Serving GPRS Support Node by sending a Wireless LAN device deactivation response to the SGSN over the UMTS network.

3.2   Handover from UMTS to Wireless LAN

Looking at the other direction of the vertical handover, namely from UMTS to Wireless LAN as shown in Fig. 16 in the Appendix A, we recognize the same three steps that are necessary for the vertical handover but the roles of the handover source and target are inverted. This leads to a similar protocol flow which differs only in network technology dependent elements. Again, the protocol flow is examined in three parts.

Wireless LAN connection establishment

When the Serving GPRS Support Node decides that an ME should perform a vertical handover from the UMTS network to the Wireless LAN, it transmits a Wireless LAN device activation request, including the information which Wireless LAN Access Point has to be used, over the existent UMTS connection. Fig. 6 visualizes this. By transferring this information, the ME does not need to scan all channels and look at all network names which are called *Service Set Identifier* (SSID) to find the appropriate Wireless LAN Access Point. At the same time, a Wireless LAN Access Point ACL entry insert request is sent to the Wireless LAN Access Point which updates its access control list. This allows the Wireless LAN Access Point to grant access for the ME without explicitly asking the Serving GPRS Support Node. If the ME receives the Wireless LAN device activation request, it associates with the Wireless LAN Access Point with a typical three way handshake. Therefore, it sends a Wireless LAN association request to the Wireless LAN Access Point, which answers with a Wireless LAN association response. This message is confirmed by the ME with an association acknowledgment and the ME is connected to the Wireless LAN Access Point over the air interface.

To be allowed to send data packets over the UMTS core network, the ME has to ask the SGSN for a packet data protocol context for every single quality of service class.

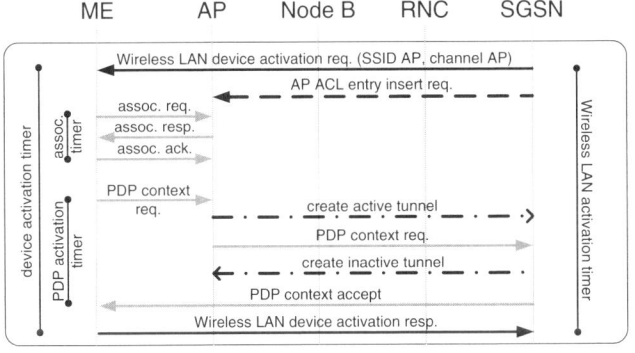

Fig. 6. Wireless LAN connection establishment

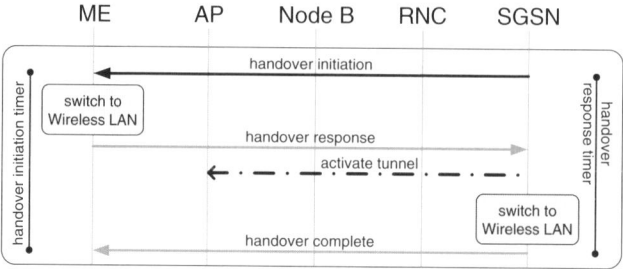

Fig. 7. Network switching from UMTS to Wireless LAN

However, the Wireless LAN Access Point has to create an IP-over-IP data tunnel, so we send the PDP context request to the Wireless LAN Access Point which processes it and creates an active IP-over-IP data tunnel from itself to the Serving GPRS Support Node. Now, the Wireless LAN Access Point requests a PDP context in the name of the ME at the SGSN. When the Serving GPRS Support Node receives this request, it establishes an inactive IP-over-IP data tunnel with the Wireless LAN Access Point as the endpoint. When this tunnel exists, the PDP context can be confirmed by sending a PDP context accept message from the SGSN to the ME over the Wireless LAN. The arrival of this message indicates the ME that one quality of service class is ready to send data packets. If all active quality of service classes in the UMTS network have been activated in the Wireless LAN, the ME has established all necessary connections and transmits the Wireless LAN device activation response to the Serving GPRS Support Node over the UMTS network. Receiving this message, the Serving GPRS Support Node knows that the Wireless LAN connection establishment is finished successfully and the actual handover can be performed.

Vertical handover - network switching from UMTS to Wireless LAN
Currently, we are connected via both interfaces and the data traffic is sent over UMTS but should be switched to the Wireless LAN. Therefore, like shown in Fig. 7, the Serving GPRS Support Node sends a handover initiation message to the ME via the UMTS network. If this message is received at the ME, it switches the device used for outgoing data connection to Wireless LAN and accepts incoming traffic from both devices. After that, the ME transmits a handover initiation response to the Serving GPRS Support Node. This message tells the SGSN that the ME has already switched and is accepting data traffic over Wireless LAN. Therefore, the Serving GPRS Support Node activates the tunnel to the Wireless LAN Access Point and is now able to transmit all further outgoing data packets to the Wireless LAN Access Point instead of the Radio Network Controller. To complete the switching process, the Serving GPRS Support Node informs the ME of the successful switching by sending a handover complete message. Doing this will finish the network switching process which means that the complete data traffic is now transmitted over the Wireless LAN and the UMTS connection can be released.

UMTS connection release

We can see the protocol details of the UMTS connection release in Fig. 8. The UMTS connection has to be released when the data traffic is completely transferred over to the Wireless LAN. Therefore, the Serving GPRS Support Node sets the IP-over-IP data tunnel to the Radio Network Controller into standby state and transmits a radio bearer release request to the RNC. After receiving the message, the RNC has to free the links to the ME and the NodeB. First, the radio bearer release request is sent from the RNC to the ME, which itself answers with a radio bearer release complete to stop the radio bearer, the service provided by the layer 2 for transfer of user data between a ME and the UTRAN. Additionally, the packet data tunnel from the Radio Network Controller to the SGSN is removed. The second step is to delete the old radio link from the active set of the NodeB and that is done by sending a radio link delete request from the Radio Network Controller to the NodeB. After deleting the radio link from the active set, a radio link delete response is sent from the NodeB back to the Radio Network Controller. With the arrival of this response, the Radio Network Controller can release the radio access bearer and transmit a radio access bearer release response to the SGSN. To complete the disassociation of the ME, it is necessary to remove the already inactive data packet tunnel to the Radio Network Controller. The connection is released now and the vertical handover from the UMTS network to the Wireless LAN is completed.

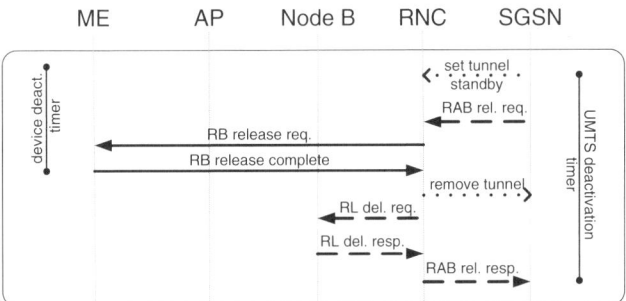

Fig. 8. UMTS connection release

4   Simulation Results

To determine the overall delay of the vertical handover procedure and to see how long the three parts of the handover process last, the handover protocol for both directions, from Wireless LAN to UMTS and UMTS to Wireless LAN, has been implemented and evaluated using the OPNET Modeler [19]. In order to evaluate the performance of the protocol without any influences of background traffic, only one ME is placed in the scenario as shown in Fig. 9. This ME moves between the coverage areas of both access technologies. It starts a voice over IP connection using the ITU-T G.711 [20] standard with an application server connected to the GGSN. This connection is needed to have an active PDP context which is taken over by the destination network. The vertical handover itself is initiated after about 200 seconds simulation time by the SGSN based on an insufficient signal strength of the ME to the currently connected network.

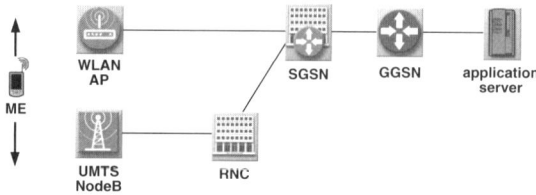

Fig. 9. Simulation scenario

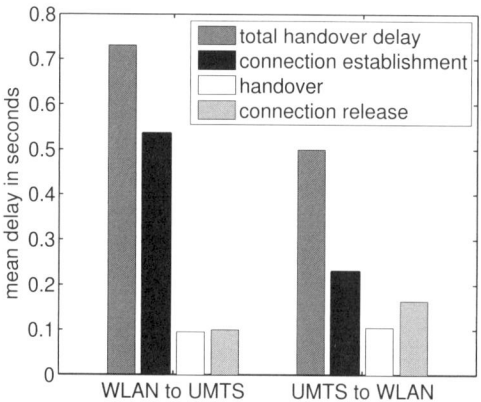

Fig. 10. Mean handover delays

The mean handover delays from 40 simulation runs are plotted in Fig. 10. In this figure and all following figures, no error bars are plotted. Due to the small variance of our simulation results, the confidence interval is in order of one millisecond. The total handover delay, indicated by the dark gray bars, is less than 750 ms for both directions and by far shorter than the Mobile IP delay with several seconds. However, included in this delay are the connection establishment and the connection release. During these two periods, packets can be transmitted normally. The period with no connection at all, the white bars, last less than 100 ms.

The figure further reveals that the connection establishment takes most of the time, especially for the Wireless LAN to UMTS handover. This results from the complex radio access bearer and tunnel setup. Unfortunately, this delay cannot be reduced and will always be around half a second for the UMTS connection setup and 230 ms for the Wireless LAN connection establishment. However, the handover delay itself, which is less than 100 ms, can be decreased which will be shown in the next subsection.

4.1 Protocol Optimization

When looking at the protocol, we recognize that the device activation complete message, the handover initiation message, and the handover response can be transmitted over both networks. Basically, the first two messages are sent over the source network,

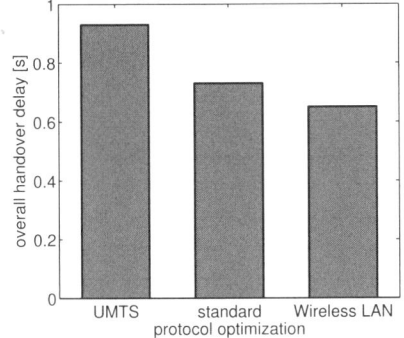

Fig. 11. Overall WLAN-UMTS handover delay

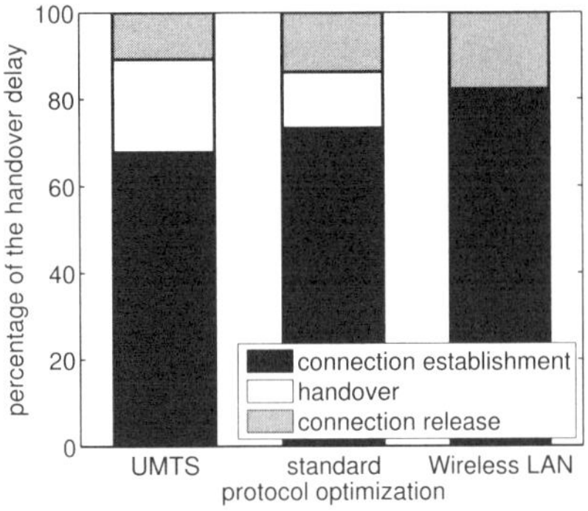

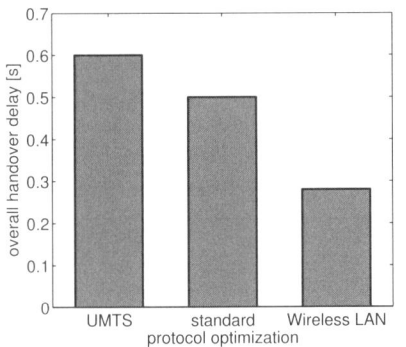

Fig. 12. Percentage of the handover parts

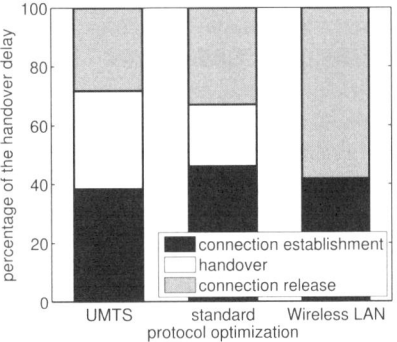

Fig. 13. Overall UMTS-WLAN handover delay

Fig. 14. Percentage of the handover parts

then the network switching of the data packets occurred at the ME and the last message is also sent over the destination network. However, due to network load, this may not be the best solution. Therefore, the vertical handover of the Mobile Equipment as well as the vertical handover of the SGSN can transmit these messages either via Wireless LAN or via UMTS. We call this parameter the network preference and we are able to define the preferred networks independently at the Mobile Equipment and the Serving GPRS Support Node. This is one possibility to optimize the vertical handover protocol performance.

In Fig. 11, the standard handover delay of 732 ms is plotted in the middle. This value includes the connection setup, the handover itself, and the Wireless LAN connection release. Transmitting the three handover messages over Wireless LAN reduces the delay to only 650 ms. If however, the load in the Wireless LAN cell is high and the three messages are transmitted using the UMTS network, the vertical handover delay increases to 930 ms.

Furthermore, in Fig. 12 the ratio of the connection establishment, the vertical handover, and the connection release of the standard protocol against the optimized two protocols is compared. In this figure, the whole handover delay of one protocol is broken down into the three parts and their ratio to the whole vertical handover delay is shown. On the x-axis the different optimizations are assigned and the y-axis shows the percentage of the connection establishment, the network switching, and the connection release. It can be noticed that switching the network technology takes about 22 percent of the total handover delay in case of a favored use of the UMTS network and does not matter in case of the Wireless LAN optimization. Therefore, it seems that the Wireless LAN optimization is particularly handling the network switching in a more efficient way.

The delay for the vertical handover from UMTS to Wireless LAN using the standard protocol is 500 ms as shown in the middle of Fig. 13. On the left side, the delay of the UMTS optimized vertical handover is shown which lasts 600 ms while the Wireless LAN optimized vertical handover lasts 280 ms. Looking at the bars in Fig. 14, it can be seen that the network switching itself takes 35 percent of the total handover delay when transmitting the handover messages over UMTS compared to a few milliseconds when they are transmitted using the Wireless LAN connection. Therefore, it does not make any sense to transmit the messages over UMTS when performing a vertical handover from UMTS to Wireless LAN.

Besides the transmission of handover messages using the faster network, another optimization, particularly for non-realtime traffic, is to forward the queued packets of the old network connection instead of destroying them when disconnecting the device. Detailed simulations of the performance such a the queue forwarding is part of future work. The last optimization can be made by delaying the disconnection and the switching of the receiving device for a while and not to switch and disconnect immediately when the handover complete message arrives. This allows to receive packets that are still on their way in the access network without resubmitting them.

5   Conclusion

In this paper, a new vertical handover approach for the integration of IP based wireless access networks into third-generation cellular networks was presented. Without loss of generality, our studies remained on the two most common technologies: UMTS as a representative of the 3G cellular networks and Wireless LAN based on the IEEE 802.11g standard as a widely used access technology. However, with a few network specific changes, it is possible to use this approach for any other IP based access network.

The focus was to develop a seamless vertical handover protocol, which means that all existing connections of the user will continue without disruption. Several approaches for a vertical handover based on Mobile IP, on the SIP, on the IMS, and on the upcoming standard IEEE 802.21 have been studied but none of them was fast enough to support real-time applications. Therefore, we developed a seamless vertical handover protocol with a low handover delay, which is network initiated and considers network knowledge as well as it handles authentication, access control, and authorization. It is designed for UMTS network operators to integrate Wireless LAN into their network to achieve a high service quality.

Our vertical handover process is divided into three parts: the connection establishment, the switching of the networks, and the connection release. Simulation results have shown that the connection establishment takes most of the handover time, about half a second for the UMTS and 230 milliseconds for the Wireless LAN connection setup. However, during this phase, data can still be sent using the old network connection. Only during the second part of the handover, the switching of the networks, no data communication is possible. Therefore, we call this period the blackout time. The first results have shown that this blackout time is with about 100 ms for every handover direction really short. However, using only the Wireless LAN connection for the message exchange reduces the blackout time to only a few milliseconds which is completely sufficient to support realtime applications. Finally, the vertical handover is completed with the connection release which takes between 100 and 170 ms depending on the technology. Concluding the results, we have shown that the complete vertical handover from Wireless LAN to UMTS ranges from 630 to 950 milliseconds according to the used optimization and the vertical handover from UMTS to Wireless LAN varies between 280 and 600 milliseconds.

Acknowledgments

The authors would like to thank OPNET Technologies Inc. for providing software license to carry out the simulations of this research.

References

1. Márquez, F.G., Rodríguez, M.G., Valladares, T.R., Miguel, T.D., Ángel Galindo, L.: Interworking of IP multimedia core networks between 3GPP and WLAN. IEEE Wireless Communications 12(3), 58–65 (2005)
2. Wu, W., Banerjee, N., Basu, K., Das, S.K.: SIP-based vertical handoff between WWANs and WLANs. IEEE Wireless Communications 12(3), 66–72 (2005)
3. Ruggeri, G., Iera, A., Polito, S.: 802.11-based wireless-LAN and UMTS interworking: requirements, proposed solutions and open issues. Computer Networks 47(2), 151–166 (2005)
4. Perkins, C.E.: IP Mobility Support. RFC 2002 (1996),
   http://www.ietf.org/rfc/rfc2002.txt
5. Sharma, S., Baek, I., Dodia, Y.: OmniCon: A Mobile IP-Based Vertical Handoff System for Wireless LAN and GPRS Links. In: 2004 International Conference on Parallel Processing Workshops (ICPPW 2004), Montreal, Quebec, Canada, pp. 330–337 (2004)
6. Soliman, H., Castelluccia, C., Malki, K.E., Bellier, L.: Hierarchical Mobile IPv6 Mobility Management (HMIPv6). RFC 4140 (2005),
   http://www.ietf.org/rfc/rfc4140.txt
7. Pries, R., Mäder, A., Staehle, D., Wiesen, M.: On the Performance of Mobile IP in Wireless LAN Environments. In: García-Vidal, J., Cerdà-Alabern, L. (eds.) Euro-NGI 2007. LNCS, vol. 4396, pp. 155–170. Springer, Heidelberg (2007)
8. Kwon, H., Cheon, K.Y., Park, A.: Analysis of WLAN to UMTS Handover. In: IEEE 66th Vehicular Technology Conference, 2007. VTC-2007 Fall. 2007, Baltimore, MD, USA, pp. 184–188 (2007)
9. Johnson, D., Perkins, C.E., Arkko, J.: Mobility Support in IPv6. RFC 3775 (2004),
   http://www.ietf.org/rfc/rfc3775.txt

10. Garroppo, R.G., Giordano, S., Lucetti, S., Risi, G., Tavanti, L.: An Experimental Cross-Layer Approach to Improve the Vertical Handover Procedure in Heterogeneous Wireless Networks. Journal of Communications Software and Systems (JCOMSS) 2(1), 40–50 (2006)

11. 3rd Generation Partnership Project.: 3GPP TS 23.228 V8.3.0.; Technical Specification Group Services and System Aspects; IP Multimedia Subsystem (IMS); Stage 2 (Release 8), V8.3.0 (2007)

12. Munasinghe, K.S., Jamalipour, A., Vucetic, B.: Interworking between WLAN and 3G Cellular Networks: An IMS Based Architecture. In: IEEE Australian Conference on Wireless Broadband and Ultra Wideband Communications (AusWireless 2006), Sydney, Australia (2006)

13. Shenoy, N., Montalvo, R.: A framework for seamless roaming across cellular and wireless local area networks. IEEE Wireless Communications 12(3), 50–57 (2005)

14. Ma, L., Yu, F., Leung, V.C.M., Randhawa, T.: A New Method to Support UMTS/WLAN Vertical Handover Using SCTP. IEEE Wireless Communications 11(4), 44–51 (2004)

15. Pries, R., Mäder, A., Staehle, D.: A Network Architecture for a Policy-Based Handover Across Heterogeneous Networks. In: OPNETWORK 2006, Washington DC, USA (2006)

16. IEEE Std. 802.11g-2003: Part 11: Wireless LAN Medium Access Control (MAC) and Physical Layer (PHY) specifications - Amendment 4: Further Higher Data Rate Extension in the 2.4 GHz Band (2003) IEEE 802.11g-2003

17. IEEE 802.21/D8.0, Vivek G. Gupta: IEEE 802.21 Standard and Metropolitan Area Networks: Media Independent Handover Services, Draft IEEE 802.21/D8.0 (2007)

18. 3rd Generation Partnership Project.: 3GPP TS 23.003 V7.6.0.; Technical Specification Group Core Network and Terminals; Numbering, addressing and identification (Release 7), V7.6.0 (2007)

19. OPNET Modeler, OPNET University Program,
    http://www.opnet.com/services/university/

20. ITU-T Recommendation G.711: Pulse Code Modulation (PCM) of Voice Frequencies. International Telecommunication Union (1998)

A   Appendix

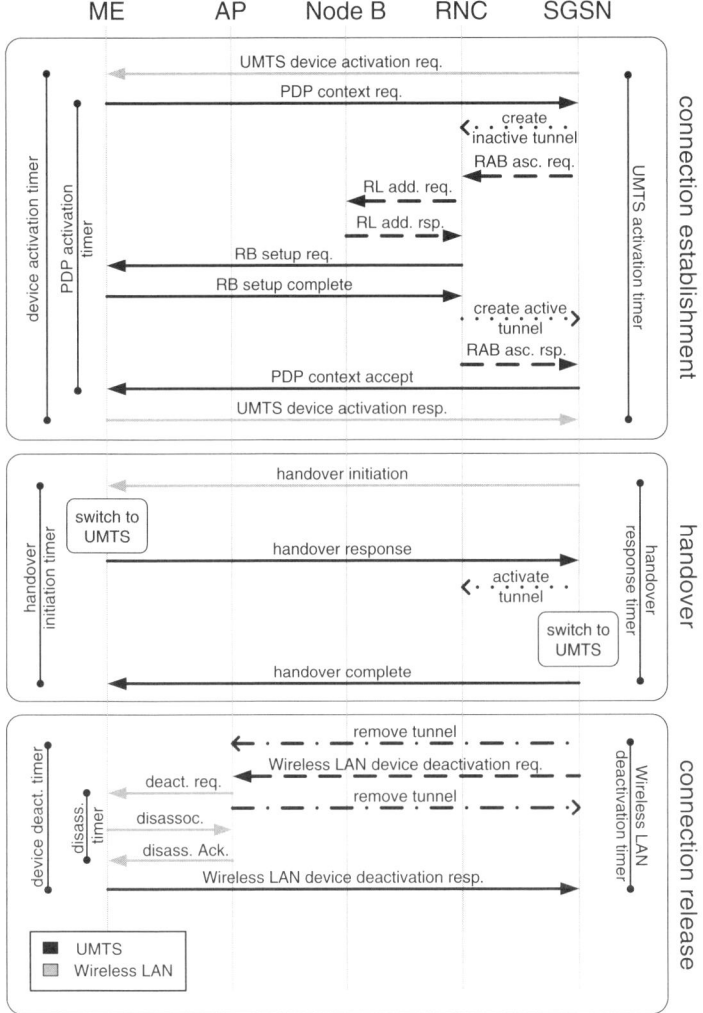

Fig. 15. Protocol for a Wireless LAN to UMTS handover

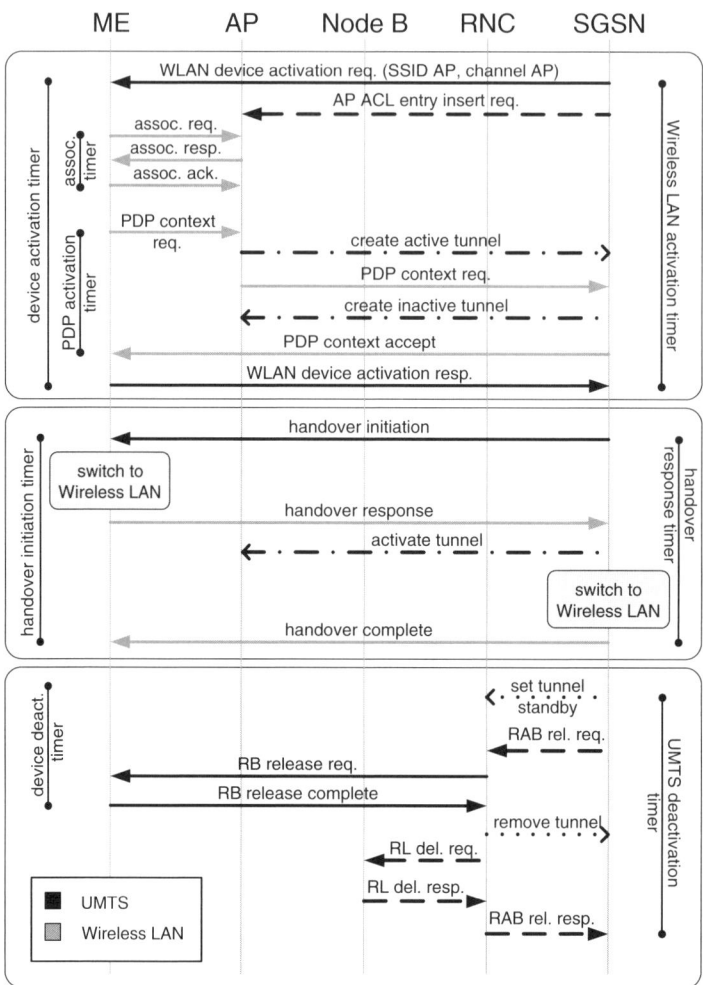

Fig. 16. Handover from UMTS to Wireless LAN

Location Management Based on the Mobility Patterns of Mobile Users

Ignacio Martinez-Arrue, Pablo Garcia-Escalle, and Vicente Casares-Giner

GIRBA Group-Instituto ITACA, Universidad Politecnica de Valencia
Camino de Vera s/n, 46022 Valencia, Spain
imartinez201j@cv.gva.es, pgarciae@upvnet.upv.es, vcasares@upvnet.upv.es

Abstract. In this paper, we enhance the mobility model presented in [1], considering a wider range of the directional movement parameter (α). Both random walk and totally directional mobility patterns are modeled. This model is used as an input to study and compare the location management cost of the movement-based strategies [2] and the distance-based scheme as a function of the mobile terminal directional mobility patterns. In [2], each time the mobile terminal revisits the last cell it had contact with the fixed network, its movement-counter is increased with probability p or it is frozen (stopped) with probability q or it is reset with probability r, $(p+q+r = 1)$. We discuss the trade off between the location update cost and the terminal paging cost. The distance-based strategy outperforms all the movement-based schemes. Among all the movement-based mechanisms, the $(p, q, r) = (0, 0, 1)$ –reset strategy– provides the best performance. All studied schemes tend to perform equally as the movement is more directional.

1 Introduction

Mobility models play an important role in the dimensioning of mobile access networks. In particular, mobility tracking is dependent of the mobility patterns of the Mobile Terminal (MT). One of the most extended mobility models used in the evaluation of mobility tracking procedures is the random walk mobility model, [3]. That model was initially considered for macrocellular scenarios. However, in microcellular environments the mobility rate is higher and MTs may move according to directional mobility patterns as the random waypoint mobility model reflects, [4].

Motivated by those facts, in this paper we enhance the quite simple and versatile mobility model presented in [1]. It generalizes the conventional random walk mobility model. In our enhanced model, we have considered a new range for the directional movement parameter, α, featuring the movement directivity [5,6]. For $0 \leq \alpha < 1$, the MT is willing to stay in the same registration area (RA) [7] where it is roaming. For $\alpha = 1$, we get a random walk model. And finally, for $\alpha > 1$, the MT is willing to leave the RA where it is roaming. Thus, both random mobility patterns (macrocellular scenarios) and directional mobility patterns (microcellular scenarios) are modeled.

L. Cerdà-Alabern (Ed.): Wireless and Mobility 2008, LNCS 5122, pp. 185–200, 2008.

In the forthcoming analysis, this mobility model is used to evaluate the impact of the users mobility patterns in location management procedures. Namely, the movement-based schemes described in [2] and the distance-based strategy have been studied.

The paper is organized as follows. Section 2 overviews the location management procedures. Section 3 describes the considered scenario and the proposed mobility model. In Sect. 4, we compute the location management costs of several movement-based schemes using our model. Section 5 deals about the evaluation of the distance-based strategy using directional mobility patterns. Some numerical results are discussed and detailed in Sect. 6. Finally, Sect. 7 draws the conclusions on the work presented in this paper.

2   Overview on Location Management

Location management is defined as the set of procedures that allow an MT being located at any time, any where, so that incoming calls may be delivered to that MT. This set consists of location update (LU) and call delivery (CD). The LU process consists of maintaining the MT location information up to date in the system databases. The database entry of an MT is updated whenever the MT triggers an LU message or an incoming call is delivered. The CD procedure is decomposed into two steps: interrogation and terminal paging (PG). Firstly, in the interrogation step, the system databases are queried to obtain the RA where the MT had contact with the fixed network (FN) for the last time. Afterwards, by means of the PG procedure, the MT is searched by polling the set of cells of the RA.

There is a trade off between LU and PG procedures. As the number of LU messages increases, the LU cost becomes higher but the PG cost decreases as the MT position is known more accurately. On the other hand, the lower is the number of LUs, the lower is the LU cost; but the uncertainty of the MT position is higher and the PG cost increases.

The LU procedures can be classified into static or dynamic strategies. In static schemes, the whole coverage area is divided into several fixed RAs called location areas (LAs). Each LA is a fixed group of neighboring cells, so that the MT delivers an LU message whenever it crosses an LA border. In dynamic strategies, the RA borders depend on the cell where the MT position was updated for the last time. In [8,9], three dynamic LU schemes were proposed. LU messages are delivered according to the elapsed time, the number of performed movements or the traveled distance by the MT from the cell where its position was updated for the last time. It is pointed out that the distance-based scheme outperforms both movement-based and time-based mechanisms. Besides, the movement-based strategy achieves a lower cost than the time-based scheme. However, the distance-based policy is more complex to be implemented because it requires to compute the distance traveled by the MT (measured in cells) using specific algorithms, [10]. Hence, the movement-based strategy may be more suitable for its simplicity. Moreover, some simple enhanced versions of the movement-based scheme, [2], outperform the classical movement policy. In [2], each time the MT revisits the last cell it had contact with the FN, its movement-counter is increased with one unit with probability p, or it is frozen (stopped) with probability q or it is reset with probability r $(p + q + r = 1)$.

Once an LU scheme has been chosen, the PG process can be carried out according to two different policies: non-selective or selective. In non-selective PG, hereinafter called one-step PG, all cells in the RA are polled simultaneously. A high PG cost is obtained but a minimum delay on the CD is achieved. In selective PG strategies [11,12], the RA is divided into several PG areas (PAs). The PAs are polled sequentially so that PAs where the called MT is more likely to be located are polled first. This policy yields a lower cost than one-step PG even though a greater delay on the CD than one-step PG is obtained.

The performance of location management procedures is dependent of the mobility patterns of the MT.

3   Proposed Mobility Model

3.1   Cell Dwell Time Distribution

An MT is roaming within a cell during a random time which is featured through a generalized gamma distribution [13], with probability density function (pdf) denoted by $f_c(t)$ and mean cell dwell time value equal to $1/\lambda_m$. Thus λ_m is the mobility rate of the MT. Besides, we denote by $f_c^*(s)$ the Laplace Transform (LT) of $f_c(t)$. Let $f_{cr}(t)$ be the pdf of the residual cell sojourn time, [14], and $f_{cr}^*(s)$ its respective LT. Both $f_c(t)$ and $f_{cr}(t)$ are related through their LTs as

$$f_{cr}^*(s) = \lambda_m \frac{1 - f_c^*(s)}{s} \tag{1}$$

3.2   Scenario and Transition Probabilities

A regular cell layout scenario with all hexagonal cells has been chosen. All cells have the same size. Figure 1 shows that each cell is surrounded by rings of cells. The innermost ring (ring 0) consists of only cell $(0, 0)$. This is the cell where the MT had its last contact with the FN. Ring 0 is surrounded by ring 1, ring 1 is surrounded by ring 2, and so on. The number of cells in ring m, denoted by $g(m)$, is given by

$$g(m) = \begin{cases} 1; & m = 0 \\ 6m; & m > 0 \end{cases} \tag{2}$$

In a random walk model, when the MT leaves its current cell, it moves to one of its six neighboring cells with probability $1/6$. In the proposed model, a single parameter denoted by α provides directivity to the MT motion. To that end, cells are labeled as it is shown in Fig. 1. Cells are divided into two sets. In the first set, cells limit with three neighboring cells in their contiguous outer ring. Such cells are located in the corners of their ring and they are labeled as $(x, 0)$, for $x \geq 1$. In the second set, cells are bordered by two neighboring cells in

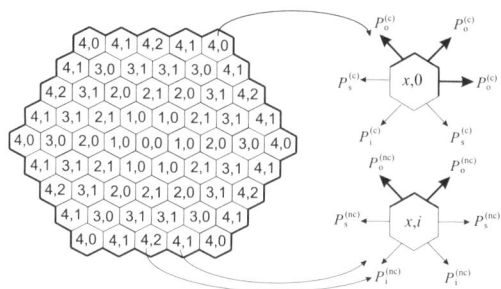

Fig. 1. Cell labeling and transition probabilities between two cells

the contiguous outer ring. They are located in the edges of their ring and they are labeled as (x, i), for $x \geq 2$ and $1 \leq i \leq \lfloor \frac{x}{2} \rfloor$. Notice ring m, for $m > 0$, is composed by 6 cells from the first set –label $(x, 0)$– and $6(m - 1)$ cells from the second set.

The transition probabilities from one cell to another are defined according to the motion of a given MT implying a movement to the outer ring (P_o), a movement to the same ring (P_s) or a movement to the inner ring (P_i).

Then, for cells of type $(x, 0)$, with $x = 1, 2, ...$, that are located in the corners (c) of the ring, we define

$$
\begin{aligned}
P_o^{(c)} &= \frac{\alpha}{3(1+\alpha)}; && \text{three edges per cell} \\
P_s^{(c)} &= \frac{1}{3(1+\alpha)}; && \text{two edges per cell} \\
P_i^{(c)} &= \frac{1}{3(1+\alpha)}; && \text{one edge per cell}
\end{aligned} \tag{3}
$$

and for cells of type (x, i), with $x = 1, 2, ...$ and $i = 1, 2, ..., \lfloor \frac{x}{2} \rfloor$, which are not placed in the corners (nc) of the ring, we define

$$
\begin{aligned}
P_o^{(nc)} &= \frac{\alpha}{2(2+\alpha)}; && \text{two edges per cell} \\
P_s^{(nc)} &= \frac{1}{2(2+\alpha)}; && \text{two edges per cell} \\
P_i^{(nc)} &= \frac{1}{2(2+\alpha)}; && \text{two edges per cell}
\end{aligned} \tag{4}
$$

Notice the above probabilities fulfill the following conditions respectively (see Fig. 1),

$$
\begin{aligned}
3P_o^{(c)} + 2P_s^{(c)} + P_i^{(c)} &= 1 \\
2P_o^{(nc)} + 2P_s^{(nc)} + 2P_i^{(nc)} &= 1
\end{aligned}
$$

The parameter α is defined as a directional movement parameter, that may take values within $[0, \infty[$. When $\alpha = 1$ we get a random walk model. As α increases from 1 to infinity, P_o grows. Hence, movements towards the outer ring from the current one become more probable –a more directional movement is modeled–. In the limit case of α tending to infinity, all cell boundary crossings would imply a movement towards the outer ring. If α decreases from 1 to 0, P_s

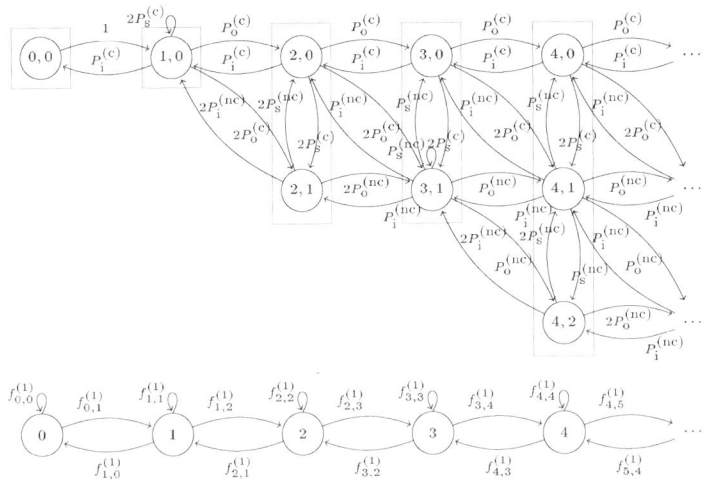

Fig. 2. Two-dimensional and one-dimensional Markov chains

and P_i arise. When $\alpha = 0$, the MT is always roaming within the same ring or it moves towards an inner ring from the current one after a cell boundary crossing.

From the previous transition probabilities, a two-dimensional (2D) Markov chain arises with states denoted by (x, i) (see Fig. 2). The label x represents the number of ring where the MT is roaming, whereas the label i indicates the type of cell where the MT is located.

For sake of simplicity in the forthcoming analysis in Sects. 4 and 5, a one-dimensional (1D) Markov chain has been derived from the previous one, as it is a common assumption, [11,12]. For this conversion, cells are grouped by rings. When this new chain is in state m, the MT is roaming in a cell within ring m. We denote by $f_{m,n}^{(1)}$ the probability that the MT moves from a cell in ring m to a cell in ring n after a single cell boundary crossing. These one-step probabilities are given by, for ring $m = 0$,

$$f_{0,1}^{(1)} = 1; \quad f_{0,0}^{(1)} = 0 \tag{5}$$

and for ring $m > 0$,

$$
\begin{aligned}
f_{m,m+1}^{(1)} &= 6\tfrac{1}{6m}3P_o^{(c)} + 6(m-1)\tfrac{1}{6m}2P_o^{(nc)} = \tfrac{1}{m}\left(\tfrac{\alpha}{1+\alpha}\right) + \tfrac{m-1}{m}\left(\tfrac{\alpha}{2+\alpha}\right) \\
f_{m,m}^{(1)} &= 6\tfrac{1}{6m}2P_s^{(c)} + 6(m-1)\tfrac{1}{6m}2P_s^{(nc)} = \tfrac{1}{m}\left(\tfrac{2}{3(1+\alpha)}\right) + \tfrac{m-1}{m}\left(\tfrac{1}{2+\alpha}\right) \\
f_{m,m-1}^{(1)} &= 6\tfrac{1}{6m}P_i^{(c)} + 6(m-1)\tfrac{1}{6m}2P_i^{(nc)} = \tfrac{1}{m}\left(\tfrac{1}{3(1+\alpha)}\right) + \tfrac{m-1}{m}\left(\tfrac{1}{2+\alpha}\right)
\end{aligned} \tag{6}
$$

All the previous probabilities depend on α and they are used to compute the LU costs and the selective PG cost.

4   Movement-Based Location Management Procedure

4.1   Description

Our mobility model has been used to evaluate our location management procedure in [2]. Basically, each MT counts the number of visits to cells. When the movement-counter reaches a defined threshold, denoted by D, the MT triggers an LU message to the FN. However, each time the MT revisits the last cell it had contact with the FN, its movement-counter is increased with one unit with probability p, or it is frozen (stopped) with probability q or it is reset with probability r $(p + q + r = 1)$. Notice that the proposals of [12], $(p, q, r) = (1, 0, 0)$, and [15], $(p, q, r) = (0, 0, 1)$, are particular cases of our general scheme.

4.2   Location Update Analysis

Figure 3 shows the 1D Markov chain, for the case of $D = 5$, used in our analysis, [2]. When the Markov chain is in state S_i, the MT is roaming in a cell within ring i. The probability $P_{\text{nab}}(D)$ (nab = no absorption) is the conditional probability that starting in state S_0 no absorption into state S_0 is produced after D movements, and it is given by

$$P_{\text{nab}}(D) = 1 - \sum_{n=1}^{D} f_{0,0}^{(n)} \tag{7}$$

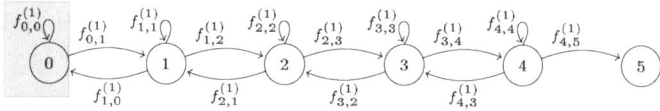

Fig. 3. State transition diagram for computing the movement-based LU cost with $D = 5$

In (7), $f_{i,j}^{(n)}$ denotes the conditional probability that state S_j is avoided at times $1, 2, ..., n-1$ and entered at time n, given that state S_i is occupied initially. As a matter of fact, $f_{0,0}^{(n)}$ can be obtained in terms of $p_{0,0}^{(n)}$ by using the following convolution equation

$$p_{0,0}^{(n)} = \sum_{k=1}^{n} f_{0,0}^{(k)} p_{0,0}^{(n-k)} \tag{8}$$

In (8), $p_{i,j}^{(n)}$ denotes the conditional probability that starting in state S_i we enter state S_j at time n (not necessarily for the first time). Obviously, $p_{i,j}^{(1)} = f_{i,j}^{(1)}$.

We denote by $M_s(z)$ the expected number of LU messages triggered by the MT in z movements, given that state S_0 is occupied initially and the value of the movement-counter of the MT is s, $s = 0, 1, ..., D-1$. It is given by

$$M_s(z) = \begin{cases} 0; & \text{for } z < D - s \\ \displaystyle\sum_{n=1}^{D-s-1} f_{0,0}^{(n)}[pM_{n+s}(z-n) + qM_{n+s-1}(z-n) + rM_0(z-n)]+ & \\ f_{0,0}^{(D-s)}[p + (p+r)M_0(z-D+s) + qM_{D-1}(z-D+s)]+ & \\ P_{nab}(D-s)[1 + M_0(z-D+s)]; & \text{for } z \geq D - s \end{cases} \quad (9)$$

Assuming that calls arrive to the MT following a Poisson process with rate λ_c, the probability $\alpha(z)$ that there are z cell boundary crossings between two call arrivals is given by (see [12]),

$$\alpha(z) = \begin{cases} 1 - (1-a)/\theta; & z = 0 \\ (1-a)^2 a^{z-1}/\theta; & z > 0 \end{cases} \quad (10)$$

In (10), $a = f_c^*(\lambda_c)$ is the LT of the MT cell dwell time (see Sect. 3.1) evaluated at $s = \lambda_c$. It represents the probability that the MT leaves its current cell before a new incoming call is received, [7]. The ratio between the call arrival rate, λ_c, and the mobility rate, λ_m, is the so called call-to-mobility ratio (CMR) [12], denoted as $\theta = \lambda_c/\lambda_m$.

Therefore, the LU cost is given by

$$C_{\text{LU}} = U\sum_{z=D}^{\infty} \alpha(z)M_0(z) = U\frac{(1-a)^2}{\theta a}M_0^*(a) \quad (11)$$

where $M_s^*(a) = \sum_{z=D-s}^{\infty} a^z M_s(z)$ and U is the cost of triggering a single LU message. The $M_0^*(a)$ term calculation is explained in Appendix A and it is completely detailed in [2].

4.3   Terminal Paging Analysis

We have considered a shortest-distance-first (SDF) partitioning scheme, [12]. The RA of the MT is partitioned into $l = \min(\eta, D)$ PAs, where η is the maximum allowable PG delay. The jth PA is denoted by A_j, where $0 \leq j < l$. Each PA contains one or more rings of cells. The PA A_j contains rings s_j to e_j where s_j and e_j are obtained as

$$s_j = \begin{cases} 0; & \text{for } j = 0 \\ \lfloor \frac{Dj}{l} \rfloor; & \text{otherwise} \end{cases} ; \quad e_j = \lfloor \frac{D(j+1)}{l} \rfloor - 1; \quad (12)$$

When an incoming call arrives to an MT, the network first determines the PAs for the called MT and then initiates the PG process. Firstly, the network simultaneously polls all cells within PA A_0. If the MT is found in A_0 the PG process ends. Otherwise, the network polls the cells in the PA A_1, and so on.

Let $\pi_{0,m}^{(s)}(z)$ ($s, m \in [0, D-1]$) be the conditional probability that starting from ring 0 (state S_0) and the initial value of the MT movement-counter is s, the MT is roaming in a cell of ring m after z movements,

$$
\pi_{0,m}^{(s)}(z) = \begin{cases} p_{0,m}^{(z)}; & \text{for } z < D-s \\ \displaystyle\sum_{n=1}^{D-s-1} f_{0,0}^{(n)}[p\pi_{0,m}^{(s+n)}(z-n)+q\pi_{0,m}^{(s+n-1)}(z-n)+r\pi_{0,m}^{(0)}(z-n)]+ & (13) \\ f_{0,0}^{(D-s)}[(p+r)\pi_{0,m}^{(0)}(z-D+s)+q\pi_{0,m}^{(D-1)}(z-D+s)]+ \\ P_{nab}(D-s)\pi_{0,m}^{(0)}(z-D+s); & \text{for } z \ge D-s \end{cases}
$$

Therefore, the probability that the MT is roaming in a cell of ring m when a call arrival occurs is given by $\pi_{0,m}^{(s)}$ evaluated at $s=0$, where

$$
\pi_{0,m}^{(s)} = \sum_{z=0}^{\infty} \alpha(z)\pi_{0,m}^{(s)}(z); \quad \text{for } 0 \le s \le D-1 \tag{14}
$$

The calculation of the previous expression is reported in Appendix B and its complete analysis is developed in [2].

Hence, using $\pi_{0,m} = \pi_{0,m}^{(0)}$, we can compute ρ_k, the probability that the MT is residing in the PA A_k when a call arrival occurs

$$
\rho_k = \sum_{i=s_k}^{e_k} \pi_{0,i} \tag{15}
$$

For a given SDF partitioning scheme, we compute the number of cells in the kth PA, denoted by $N(A_k)$, and the number of polled cells before the MT is successfully located (given by w_k),

$$
N(A_k) = \sum_{j=s_k}^{e_k} g(j) \tag{16}
$$

$$
w_k = \sum_{n=0}^{k} N(A_n) \tag{17}
$$

Then, the expected terminal PG cost per call arrival, denoted by C_{PG}, is

$$
C_{\text{PG}} = V\sum_{k=0}^{l-1} \rho_k w_k = V\sum_{k=0}^{l-1}\left[\sum_{i=s_k}^{e_k}\pi_{0,i}\sum_{n=0}^{k}\sum_{j=s_n}^{e_n} g(j)\right] \tag{18}
$$

where V is the cost of polling a cell.

Note that $l=1$ (a single PG step) provides a single PA that agrees with the RA. Thus the previous analysis is also valid if a one-step PG policy is used.

5   Distance-Based Location Management Procedure

5.1   Description

If a random walk model is used, it is a well known result that the distance-based mechanism outperforms the movement-based strategy for $(p,q,r) = (1,0,0)$

(see [8,9]). However, an unknown issue is how the MT movement directivity influences the distance-based scheme cost. Then, we study the distance-based strategy by using our proposed mobility model as it is done in [1].

In the distance-based scheme, the distance between the MT current cell and the cell where it had contact with the FN for the last time is measured in terms of cells after each cell boundary crossing. Whenever the MT reaches the distance threshold, denoted by D, it delivers an LU message.

5.2   Location Update Analysis

Figure 4 shows the 1D Markov chain that is used in our analysis. Each state of that Markov chain represents the number of ring where the MT is roaming. The MT is roaming in ring m during a random time with pdf denoted by $f_m(t)$. Let $f_m^*(s)$ be the LT of $f_m(t)$. Also, we denote by $f_{mr}(t)$ the residual sojourn time distribution in ring m. The LT of $f_{mr}(t)$ is given by $f_{mr}^*(s)$. For ring $m = 0$, obviously we have $f_m^*(s) = f_c^*(s)$ and $f_{mr}^*(s) = f_{cr}^*(s)$. For $m > 0$,

$$
\begin{aligned}
f_m^*(s) &= \frac{f_c^*(s)(1 - f_{m,m}^{(1)})}{1 - f_{m,m}^{(1)} f_c^*(s)} \\
f_{mr}^*(s) &= \frac{f_{cr}^*(s)(1 - f_{m,m}^{(1)})}{1 - f_{m,m}^{(1)} f_c^*(s)}
\end{aligned}
\tag{19}
$$

The transition probabilities of the Markov chain shown in Fig. 4 represent the probability that the MT moves forward (fw) or backward (bk) whenever it moves from one ring to another. These transition probabilities are expressed as

$$
\begin{aligned}
P_{\text{fw}}(m) &= \frac{f_{m,m+1}^{(1)}}{f_{m,m+1}^{(1)} + f_{m,m-1}^{(1)}} \\
P_{\text{bk}}(m) &= \frac{f_{m,m-1}^{(1)}}{f_{m,m+1}^{(1)} + f_{m,m-1}^{(1)}}
\end{aligned}
\tag{20}
$$

We assume that call arrivals occur following a Poisson process with call arrival rate λ_c. Then, we denote by $r_{1,r}(i, j, \lambda_c)$ the probability that the MT leaves the set composed by states i to j (with $i < j$) through state j before a new incoming call arrives, given that the MT is initially in state i. Besides, let $rr_{1,r}(i, j, \lambda_c)$ be the analogous residual probability. Also we define $r_{1,l}(i, j, \lambda_c)$ as the probability that the MT leaves the set of states from i to j (with $i < j$) through state i

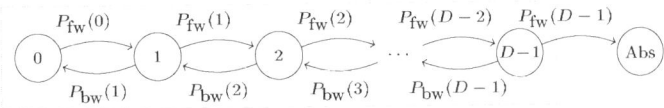

Fig. 4. One-dimensional Markov chain for computing the distance-based LU cost

before a new call arrival occurs, assuming that the MT starts in state i. Those probabilities may be computed through the recursive method proposed in [16],

$$
\begin{aligned}
rr_{1,r}^*(i,j,\lambda_c) &= P_{\text{fw}}(i)f_{ir}^*(\lambda_c)\left(r_{1,r}^*(i+1,j,\lambda_c)+r_{1,1}^*(i+1,j,\lambda_c)r_{1,r}^*(i,j,\lambda_c)\right)\\
r_{1,r}^*(i,j,\lambda_c) &= \frac{P_{\text{fw}}(i)f_i^*(\lambda_c)r_{1,r}^*(i+1,j,\lambda_c)}{1-P_{\text{fw}}(i)f_i^*(\lambda_c)r_{1,1}^*(i+1,j,\lambda_c)}\\
r_{1,1}^*(i,j,\lambda_c) &= \frac{P_{\text{bk}}(i)f_i^*(\lambda_c)}{1-P_{\text{fw}}(i)f_i^*(\lambda_c)r_{1,1}^*(i+1,j,\lambda_c)}
\end{aligned}
\tag{21}
$$

where the initial conditions are given by

$$
\begin{aligned}
r_{1,r}^*(j,j,\lambda_c) &= P_{\text{fw}}(j)f_j^*(\lambda_c)\\
r_{1,1}^*(j,j,\lambda_c) &= P_{\text{bk}}(j)f_j^*(\lambda_c)
\end{aligned}
$$

We denote by n the number of LU messages triggered by the MT between two call arrivals. From (21), the probability that n LU messages are delivered between two incoming calls, $P(n \text{ LUs})$, can be obtained as

$$
P(n\,\text{LUs})=\begin{cases}1-rr_{1,r}^*(0,D-1,\lambda_c); & n=0\\ rr_{1,r}^*(0,D-1,\lambda_c)[r_{1,r}^*(0,D-1,\lambda_c)]^{n-1}[1-r_{1,r}^*(0,D-1,\lambda_c)]; & n>0\end{cases}
\tag{22}
$$

Then, the LU cost per call arrival is given by

$$
C_{\text{LU}} = U\sum_{i=0}^{\infty} iP(i\,\text{LUs}) = U\frac{rr_{1,r}^*(0,D-1,\lambda_c)}{1-r_{1,r}^*(0,D-1,\lambda_c)}
\tag{23}
$$

In (23), U is the cost of triggering a single LU message.

5.3   Terminal Paging Analysis

We have considered an SDF PG policy as it was described in Sect. 4.3. Thus, the RA is divided into $l = \min(\eta, D)$, where η is the maximum allowable PG delay. We denote by A_j the jth PA, reminding that $0 \le j < l$. The PA A_j contains rings s_j to e_j where s_j and e_j are obtained according to (12).

Assuming the MT always starts in state 0 in a distance-based scheme, let $\pi_{0,m}(n,k)$ denote the probability that the MT is roaming in a cell of ring m when a new incoming call arrives, given that n LU messages have been triggered before and the MT has left and come back to ring m for k times since the last LU, $\pi_{0,m}(n,k)$

$$
=\begin{cases}
rr_{1,r}^*(0,m-1,\lambda_c)[1-f_m^*(\lambda_c)][f_m^*(\lambda_c)]^k\\
\left[P_{\text{fw}}(m)r_{1,1}^*(m+1,D-1,\lambda_c)+P_{\text{bk}}(m)r_{r,r}^*(0,m-1,\lambda_c)\right]^k; & n=0;\ k\ge 0\\
rr_{1,r}^*(0,D-1,\lambda_c)r_{1,r}^*(0,D-1,\lambda_c)^{n-1}\\
r_{1,r}^*(0,m-1,\lambda_c)[1-f_m^*(\lambda_c)][f_m^*(\lambda_c)]^k\\
\left[P_{\text{fw}}(m)r_{1,1}^*(m+1,D-1,\lambda_c)+P_{\text{bk}}(m)r_{r,r}^*(0,m-1,\lambda_c)\right]^k; & n>0;\ k\ge 0
\end{cases}
\tag{24}
$$

We denote by $\pi_{0,m}(n)$ the probability that the MT is roaming in ring m when a call arrival occurs, given that n LUs were performed since the last call arrival. It is obtained as

$$\pi_{0,m}(n) = \sum_{k=0}^{\infty} \pi_{0,m}(n,k)$$

$$= \begin{cases} \dfrac{rr_{1,r}^*(0, m-1, \lambda_c)[1 - f_m^*(\lambda_c)]}{1 - f_m^*(\lambda_c)[P_{\text{fw}}(m)r_{1,1}^*(m+1, D-1, \lambda_c) + P_{\text{bk}}(m)r_{r,r}^*(0, m-1, \lambda_c)]}; & n = 0 \\[3mm] \dfrac{rr_{1,r}^*(0, D-1, \lambda_c)[r_{1,r}^*(0, D-1, \lambda_c)]^{m-1}r_{1,r}^*(0, m-1, \lambda_c)[1 - f_m^*(\lambda_c)]}{1 - f_m^*(\lambda_c)[P_{\text{fw}}(m)r_{1,1}^*(m+1, D-1, \lambda_c) + P_{\text{bk}}(m)r_{r,r}^*(0, m-1, \lambda_c)]}; & n > 0 \end{cases}$$

(25)

Then, the probability $\pi_{0,m}$ that an MT is roaming in a cell of ring m when a call arrival occurs is obtained by adding the probabilities of (25),

$$\pi_{0,m} = \sum_{n=0}^{\infty} \pi_{0,m}(n)$$

$$= \frac{[1 - f_m^*(\lambda_c)][rr_{1,r}^*(0, m-1, \lambda_c)] + \dfrac{rr_{1,r}^*(0, D-1, \lambda_c)r_{1,r}^*(0, m-1, \lambda_c)}{1 - r_{1,r}^*(0, D-1, \lambda_c)}}{1 - f_m^*(\lambda_c)[P_{\text{fw}}(m)r_{1,1}^*(m+1, D-1, \lambda_c) + P_{\text{bk}}(m)r_{r,r}^*(0, m-1, \lambda_c)]}$$

(26)

Once the probabilities $\pi_{0,m}$ have been obtained, the PG cost is obtained again according to expressions (15), (16), (17), and (18) in Sect. 4.3.

6   Numerical Results and Discussion

The total location management cost $C_T = C_{\text{LU}} + C_{\text{PG}}$ has been evaluated for a set of parameters, with main emphasis in the directional movement parameter (α) and the set of probabilities p, q, and r. In all the following figures, the values $U = 10$ and $V = 1$ have been chosen as it is done in [7,12]. A high mobility environment has been considered to compare the location management schemes, i.e. $\theta = 0.1$.

Firstly, we have analyzed the influence of the three parameters, p, q and r in the total location management cost C_T. In [2] it was shown that the minimum C_T is reached for $r = 1$ (hence for $p = q = 0$), when a random walk model $(\alpha = 1)$ is used. We have studied the behavior of C_T for other values of α. Thus, we report Fig. 5(a) $(\alpha = 0.1)$ and Fig. 5(b) $(\alpha = 100)$. Both of them show that the set of values $(p, q, r) = (0, 0, 1)$ yields the best performance again; whereas the highest location management cost is obtained by the set of values $(p, q, r) = (1, 0, 0)$. However, it can be observed that C_T is less sensitive to p, q, and r as α increases. For $\alpha = 0.1$, it is shown in Fig. 5(a) that the difference between the minimum C_T $(r = 1)$ and the maximum C_T $(p = 1)$ is greater than 39%. On the other hand, for $\alpha = 100$ (Fig. 5(b)) the maximum C_T difference is lower than 0.2%.

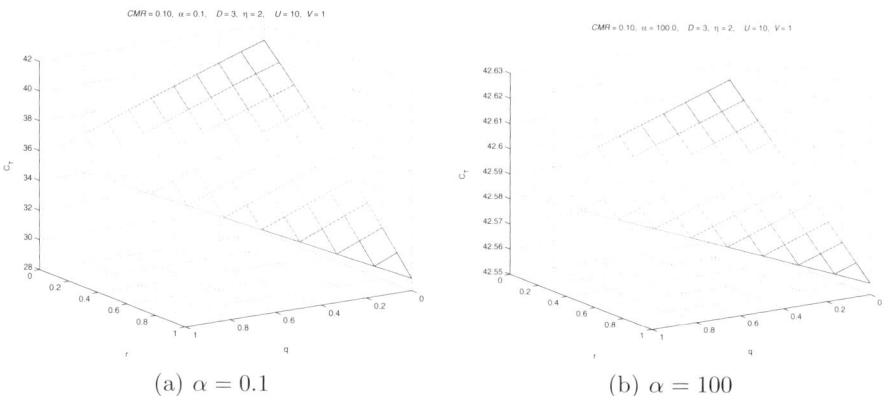

(a) $\alpha = 0.1$ (b) $\alpha = 100$

Fig. 5. Total location management cost for $D = 2$, $\eta = 2$, $U = 10$, $V = 1$ and $\theta = 0.1$

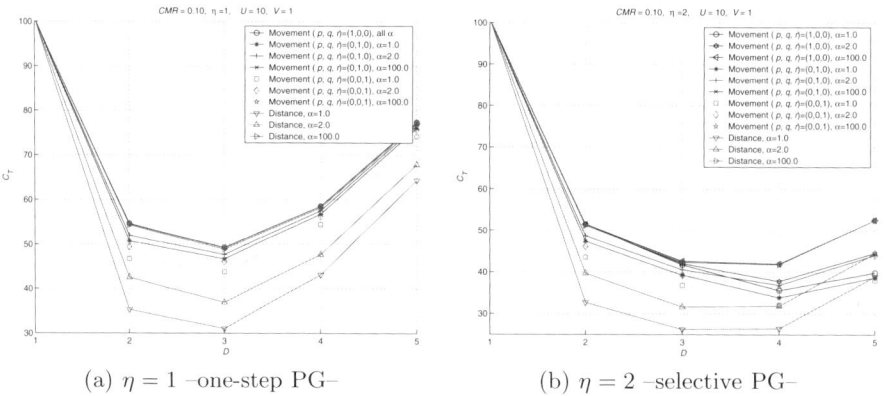

(a) $\eta = 1$ –one-step PG–   (b) $\eta = 2$ –selective PG–

Fig. 6. Total location management cost for a set of values of α, $U = 10$, $V = 1$ and $\theta = 0.1$

Next, Fig. 6 plots C_T in terms of D and $(p, q, r) = (1, 0, 0)$ –movement strategy–, $(p, q, r) = (0, 1, 0)$ –frozen strategy–, and $(p, q, r) = (0, 0, 1)$ –reset strategy–. All these strategies are compared with the distance-based scheme considering $\alpha = 1$ –random walk–, $\alpha = 2$ and $\alpha = 100$. Both one-step PG (Fig. 6(a)) and selective PG with $\eta = 2$ (Fig. 6(b)) have been considered.

All the location management costs shown in Fig. 6(a) and Fig. 6(b) are convex functions with a minimum value. That minimum value provides the optimum cost per call arrival (denoted by C_T^*). It can be obtained through adjusting the LU threshold, D. The value of D that yields C_T^* is known as the optimum threshold (that we denote by D^*). In all studied location management strategies,

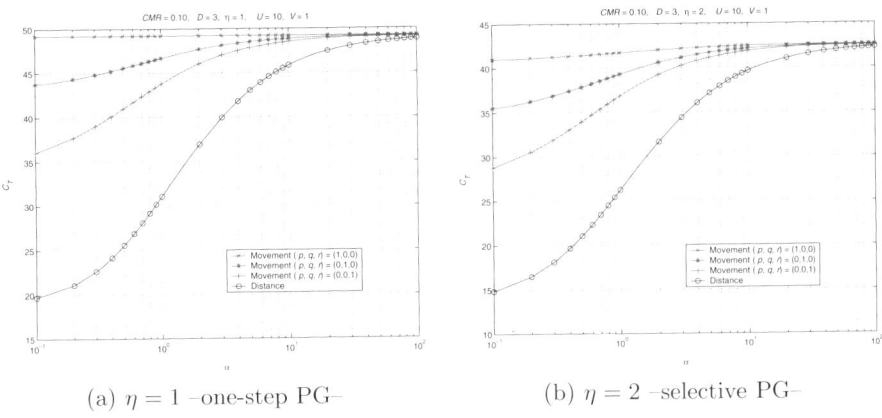

(a) $\eta = 1$ –one-step PG–   (b) $\eta = 2$ –selective PG–

Fig. 7. Total location management cost for $D = 3$, $U = 10$, $V = 1$ and $\theta = 0.1$

the PG cost function (C_{PG}) is an increasing function of D, whereas the LU cost function (C_{LU}) is a decreasing function of D. Hence, when D is greater than D^*, the PG cost dominates C_T. On the other hand, if D is lower than D^*, the LU cost dominates C_T.

Depending on the used PG policy, C_T varies. If a selective PG mechanism is employed, C_{PG} is lower than the one-step C_{PG}. This fact leads to achieve a lower value of C_T^* when selective PG is chosen. In this case, the C_{PG} dominates C_T for higher values of D. Therefore, a greater value of D^* is obtained as it is shown in Fig. 6(b).

In Fig. 6(a) and Fig. 6(b), the distance-based strategy outperforms all the other studied strategies. However, the distance-based C_T increases quickly as α becomes higher, so that the distance-based cost is more sensitive to α than all the other studied schemes. Then, in Fig. 7(a) and Fig. 7(b) we have finally tested the influence of the parameter α in C_T for all before mentioned strategies. While Fig. 7(a) shows the case of one-step PG, Fig. 7(b) shows the case of SDF PG with $\eta = 2$. In both figures, clearly the best C_T is achieved by the distance-based strategy. Sorted from lower to higher C_T and for a given value of α, we can assert that $C_{T\ distance} < C_{T\ reset} < C_{T\ frozen} < C_{T\ movement}$.

In the above figures it can be seen that all studied strategies tend to perform equally as α increases. In the limit case of α approaching to infinity, all strategies would yield the same C_T. This fact occurs because the movement of an MT provides a high probability of performing cell boundary crossings towards an outer ring for large values of α. We can remark that the differences between the total location management costs are not significant for values of $\alpha > 10$ (less than 9%). In this case, as MTs move directionally we consider that a reset scheme $-(p, q, r) = (0, 0, 1)$– combined with selective PG would be a suitable location management solution due to its simplicity.

7   Conclusions

In this paper, an enhanced mobility model has been presented. The model can be used in microcellular environments, where a more deterministic roaming of the MTs could be expected. The mobility model has been tested to evaluate several location managements schemes, in particular, a general framework of the movement-based scheme and the distance-based strategy, which is more complex to implement. Among all the movement-based strategies, the $(p, q, r) = (0, 0, 1)$ –reset strategy– provides the best performance from the signaling load point of view. In addition to this result, we confirm the very well known fact that selective PG significantly reduces the PG signaling load on the common air interface. Therefore, the implementation of our proposal is worthy of consideration in real cellular systems as we believe it can be done in an easy manner.

Acknowledgments

This work has been financed by the *Ministerio de Educacion y Ciencia* (Spain) under project numbers TEC2004-06437-C05-01 and TSI2007-66869-C02-02.

References

1. Martinez-Arrue, I., Garcia-Escalle, P., Casares-Giner, V.: Mobile user location management under a random-directional mobility pattern for PCS networks. In: Third International Working Conference on Performance Modelling and Evaluation of Heterogeneous Networks, P28/1–P28/10 (2005)
2. Casares-Giner, V., Garcia-Escalle, P.: On Movement-based mobility tracking strategy-A general framework. In: IEEE Wireless Communications and Networking Conference, vol. 4, pp. 1957–1962 (2004)
3. Camp, T., Boleng, J., Davies, V.: A survey of mobility models for ad hoc network research. Wireless Communications and Mobile Computing 2(5), 483–502 (2002)
4. Bettstetter, C., Hartenstein, H., Perez-Costa, X.: Stochastic properties of the random waypoint mobility model. Wireless Networks 10(5), 555–567 (2004)
5. Lombardo, A., Palazzo, S., Schembra, G.: A comparison of adaptive location tracking schemes in personal communications networks. International Journal of Wireless Information Networks 7(2), 79–89 (2000)
6. Tung, T., Jamalipour, A.: Adaptive location management strategy to the distance-based location update technique for cellular networks. In: IEEE Wireless Communications and Networking Conference, vol. 1, pp. 172–176 (2004)
7. Garcia-Escalle, P., Casares-Giner, V., Mataix-Oltra, J.: Reducing location update and paging costs in a PCS network. IEEE Transactions on Wireless Communications 1(1), 200–209 (2002)
8. Bar-Noy, A., Kessler, I., Sidi, M.: Mobile users: To update or not to update? In: Proceedings of INFOCOM 1994, pp. 570–576 (1994)
9. Bar-Noy, A., Kessler, I., Sidi, M.: Mobile users: To update or not to update? Wireless Communications Journal 1(2), 175–185 (1995)
10. Vidal, R., Paradells, J., Casademont, J.: Labelling mechanism to support distance-based dynamic location updating in cellular networks. IEEE Electronics Letters 39(20) (2003)

11. Ho, J.S.M., Akyildiz, I.F.: Mobile user location update and paging under delay constraints. Wireless Networks 1(4), 413–425 (1995)
12. Akyildiz, I.F., Ho, J.S.M., Lin, Y.B.: Movement-based location update and selective paging for PCS networks. IEEE/ACM Transactions on Networking 4(4), 629–638 (1996)
13. Zonoozi, M.M., Dassanayake, P.: User mobility modeling and characterization of mobility patterns. IEEE Journal on Selected Areas in Communications 15(7), 1239–1252 (1997)
14. Kleinrock, L.: Queueing systems, vol. 1. John Wiley & Sons, Chichester (1975)
15. Casares-Giner, V., Mataix-Oltra, J.: On movement-based mobility tracking strategy-An enhanced version. IEEE Communications Letters 2(2), 45–47 (1998)
16. Casares-Giner, V.: Variable bit rate voice using hysteresis thresholds. Telecommunication Systems 17, 31–62 (2001)

A   Analytical Tools for the Location Update Cost Calculation in a Movement-Based Scheme

In the Sect. 4.2, $M_0^*(a)$ is used in the final movement-based LU cost expression obtained in (11). In order to compute this term, we define the column vector $\mathbf{M}^*(a)$ with D components

$$\mathbf{M}^*(a)^t = \left[\, M_0^*(a),\ M_1^*(a),\ \dots,\ M_{D-2}^*(a),\ M_{D-1}^*(a)\,\right] \tag{27}$$

The following system is verified

$$\left(\mathbf{I} - q\mathbf{F} - p\mathbf{G} - \mathbf{H}(r)\right)\mathbf{M}^*(a) = \mathbf{J} \tag{28}$$

where matrices $\mathbf{F}$, $\mathbf{G}$, $\mathbf{H}(r)$ and the column vector $\mathbf{J}$, are given as follows. Matrix $\mathbf{F}$, is a Toeplitz matrix,

$$\mathbf{F}_{(i,j)} = \begin{cases} a^{j-i+1} f_{0,0}^{(j-i+1)}; & 1 \le i \le j \le D \\ 0; & 1 \le j < i \le D \end{cases} \tag{29}$$

Matrix $\mathbf{G}$ is obtained by shifting cyclically one place to the right each row of $\mathbf{F}$. Then, the second column of $\mathbf{G}$ is the first column of $\mathbf{F}$, the third column of $\mathbf{G}$ is the second column of $\mathbf{F}$, and so on.

The (i,j)-entry of matrix $\mathbf{H}(r)$, for $1 \le i \le D$, is given by

$$\mathbf{H}(r)_{(i,j)} = \begin{cases} \sum_{n=1}^{D-i+1} (ra^n - a^D) f_{0,0}^{(n)} + a^{D-i+1}; & j = 0 \\ 0; & 1 < j \le D \end{cases} \tag{30}$$

Finally, the column vector $\mathbf{J}$ with D components is given by

$$(1-a)\mathbf{J}^t = \Big[\ (pf_{0,0}^{(D)} + P_{nab}(D))a^D,\ (pf_{0,0}^{(D-1)} + P_{nab}(D-1))a^{D-1},$$
$$\dots,\ (pf_{0,0}^{(2)} + P_{nab}(2))a^2,\ (pf_{0,0}^{(1)} + P_{nab}(1))a\ \Big] \tag{31}$$

B   Analytical Tools for the Terminal Paging Cost Calculation in a Movement-Based Scheme

In the Sect. 4.3, the probability $\pi_{0,m}^{(s)}$ in (14) can be obtained if we define a column vector $\boldsymbol{\pi}_m$ with D components such that

$$\boldsymbol{\pi}_m^t = \left[\ \pi_{0,m}^{(0)}\ ,\ \pi_{0,m}^{(1)}\ ,\ \dots\ ,\ \pi_{0,m}^{(D-2)}\ ,\ \pi_{0,m}^{(D-1)}\ \right] \tag{32}$$

Then we can write the following set of equations

$$\big(\mathbf{I} - q\mathbf{F} - p\mathbf{G} - \mathbf{H}(r)\big)\boldsymbol{\pi}_m = \mathbf{L}_m - \mathbf{M}_m + \mathbf{N}_m \tag{33}$$

where matrices $\mathbf{F}$, $\mathbf{G}$, and $\mathbf{H}(r)$, were given in Appendix A, and the column vectors $\mathbf{L}_m$, $\mathbf{M}_m$ and $\mathbf{N}_m$, are given by

$$\mathbf{L}_m^t = \left[\ \sum_{n=0}^{D-1} \alpha(n)p_{0,m}^{(n)}\ ,\ \sum_{n=0}^{D-2} \alpha(n)p_{0,m}^{(n)}\ ,\ \dots\ ,\ \sum_{n=0}^{1} \alpha(n)p_{0,m}^{(n)}\ ,\ \alpha(0)p_{0,m}^{(0)}\ \right] \tag{34}$$

$$\mathbf{M}_m^t = \left[\ \sum_{n=1}^{D-1} a^n f_{0,0}^{(n)} \sum_{k=0}^{D-n-1} \alpha(k)p_{0,m}^{(k)}\ ,\ \sum_{n=1}^{D-2} a^n f_{0,0}^{(n)} \sum_{k=0}^{D-n-2} \alpha(k)p_{0,m}^{(k)}\ ,\right.$$
$$\left. \dots,\quad a f_{0,0}^{(1)}\alpha(0)p_{0,m}^{(0)}\ ,\qquad\qquad 0\qquad\quad \right] \tag{35}$$

$$\mathbf{N}_m^t = \left[\ c_m(D),\ c_m(D-1),\ \dots\ ,\ c_m(2),\ c_m(1)\ \right] \tag{36}$$

where $c_m(D) = (P_{nab}(D-1))(\alpha(D) - a^D \alpha(0))p_{0,m}^{(0)}$

Performance Evaluation of Overlay-Based Range Queries in Mobile Systems

Amine M. Houyou, Alexander Stenzer, and Hermann De Meer

University of Passau, Faculty for Mathematics and Computer Science,
Chair for Computer Networks and Communication, Innstr. 43
94032 Passau, Germany
{houyou,stenzer,demeer}@fim.uni-passau.de

Abstract. Current mobility management systems are operator centralized, and focused on single link technologies. In heterogeneous wireless mesh networks, vertical handovers could be a lengthy procedure. In order to support context-aware handovers between heterogeneous wireless cells, the mobile user needs to access information managed by administratively separate domains. Handover does not occur based on blind discovery mechanisms on the link layer but based on a pre-discovery of the wireless mesh topology using a context-aware search process. The mobile user has access to an overlay which forms a middleware that connects separate management domains, while allowing the user to retrieve uniform descriptions of underlying access technologies. Based on the topology information and access capabilities of the user, a context-aware handover is carried out. The overlay queries are sent to edge management servers to retrieve the cells' status as near to real-time as possible. The query system uses the geographic context to both address the managed objects as well as structuring the overlay. An analytic study and simulation are used to quantify the communication overhead of such a range query. A design methodology is also extracted from both studies.

1 Introduction

The wireless infrastructure so far has been laid out by networks operators in order to reach their customers through investing into well-organized and hierarchically structured cellular networks. The transition between wireless cells is closely monitored and tracked by the network. The location attribute of the mobile user is described through the cell's unique identifier and is in turn stored in a hierarchical structured data-base. Visited location registries (VLR) communicate with home location registries (HLR) to update the network as a whole of the location of each mobile node in the system. A mobile node only needs to report its current cell ID, for incoming calls to be directed to it. Some predictive measures could be also taken knowing the topology and direction of the movement of a mobile node [5]. A home location registry (HLR) can also store the location updates (LUs), sent from a communicating terminal, or use paging to indicate to dormant devices their current cell IDs [2].

L. Cerdà-Alabern (Ed.): Wireless and Mobility 2008, LNCS 5122, pp. 201–219, 2008.

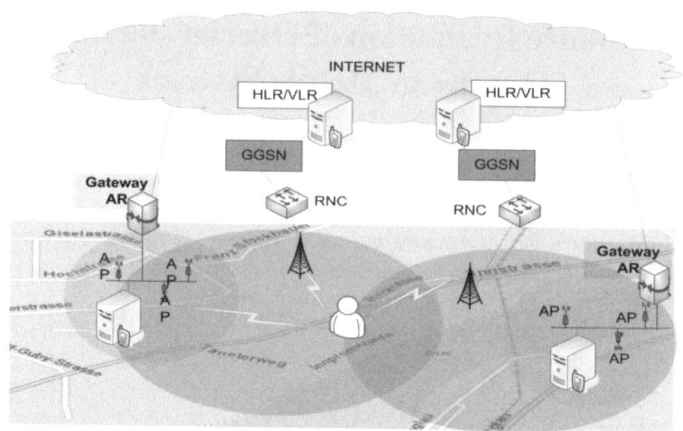

Fig. 1. The distributed nature of cellular and wireless networks management

Assuming that the mobile user can connect to any of the heterogeneous wireless mesh networks (WMN) [1] (example shown in Figure 1), the current handover process can only rely on link and network layer discovery mechanisms called beaconing. The cells and the mobile node beacon each other to discover movement. In a network centric approach, once the mobile node is located (i.e. the cell and the MN can see each other), a further assignment of resources could be done by the operator.

In the approach of this paper, the user can select out of several alternatives to which cell a handover should be started. Similar to the context-based handover advocated in [3], a mobile node could utilize the topology information to make a multiple criteria decision. This type of handover is triggered with the help of cross-layer mechanisms (see [5]).

Assuming the wireless mesh topology belongs to different domains, the problem of identifying this topology, from the mobile node's point of view, is that of querying each management domain separately for available cells. In other words, the mobile node requires a range query in a distributed and heterogeneous data base. The overlay can connect the data bases and mediate the data they manage through a single search process, which then transforms the problem to a single range query in an overlay.

The range query in overlay networks problem has two folds: first, the placement of the data items in the overlay network. Data items should be as close as possible to each other in the overlay as they are semantically. This minimizes the fragmentation of the search requests required to cover a given search or query box. The reader should be familiar with the basics of a Distributed Hash Table (DHT) structure. DHTs split a continuous key space rather equally between participating peers, similar to the amount of routing information each peer stores locally. The hash values refer to an ID space (one-dimensional in Chord [14] for instance) called key space. It is worth noting that most structured P2P protocols have been optimized to solve an exact key match routing problem (called mediation). The continuous key space normally is matched to a random application level value (or objects), which allows a certain load balancing of both the distribution of application objects among the peers and the communication or mediation overhead.

If used unchanged, the mediation effort to launch a range query would first result into a splitting of the query to retrieve each single key contained in that range. Unless the data is structured in a way that two neighbouring keys on the P2P key space refer to two semantically close data items too, the resulting mediation overhead would quickly explode. The effort to refit the overlay structure to the data ranges results in a clustering of data among neighbouring peers. The overlay routing effort could be done once for each cluster. Reaching each cluster requires the largest lookup cost in terms of overlay hops. Recursively the remaining peers could be queried along the cluster. This type of research is not new and the reader could be further referred to [4] for a more generalized study.

This paper presents a design strategy for an overlay-based location-aware distributed data base problem. The overlay deals with the mediation problem, while the location-aware range query analysis shapes the structure of the overlay network. The paper is structured as follows. The related work is presented in Section 2. Section 3 explains the application domain and original architecture requirements for supporting context aware vertical handovers. The basic overlay requirements are explained too. Section 4 presents the asymptotic analysis of the range query cost. The design strategy is extracted from the analytic study. A simulative case study is presented in Section 5. Section 6 concludes the paper.

2 Related Work

While mobility research has matured regarding cellular networks and even 3G and beyond scenarios [2], major efforts are still needed in WMNs and Ad-Hoc networks [1]. Using overlays to manage mobility or as a support for mobility has also resulted in several proposals including earlier work by Stemm and Katz [13] supporting vertical handovers. I3-based protocol [16] followed as a large scale routing framework replacing mobile IP. Similar to the home and visited location registries the Palma P2P protocol manages the user's location [12]. The work presented in this paper looks at supporting the mobile user itself in identifying the wireless diversity offered in 4G scenarios. The location management and mobility are not the focus here.

An important aspect of the proposed use of overlay is the support of location-based range queries. The mobile user, and based on its location, can query the location based service[1] [5] offered by the overlay network.

Range queries in P2P systems have addressed several issues such as non uniform partitioning of the object space. In this type of range queries, data ranges are neither continuous nor uniformly partitioned. This is the case for dictionary entries or names, where given ranges abruptly change. For this type of queries, several proposals have been made including Kademlia [7] and more systematically P-Grid based overlay [4]. P-Grids are structured as prefix hash trees (PHT) (used in traditional database research). This suggests organizing the overlay as a data trie [4], which addresses nodes according to a binary tree structure. For instance, a peer addressed with "001"

---

[1] Schiller defines Location-based services (LBS) as integrating geographic location with the general notion of a service [11].

would be on the route to all nodes or objects whose addresses share this prefix. In order to extend the scope of each peer at the lower part of the binary tree, these latter edge peers include pointer to other far off edge peers. Combined with search algorithms at the edge, the proposal in [4] is well suited for fragmented keyword based range queries. It is though worth noting that once at the edge the search algorithms are closer to flooding.

Another family of solutions is based on the multi-dimensional DHT CAN [9]. In [10] a geographic coding is proposed. Although offering a better mapping between the multiple dimensions needed to describe a geographic zone, CAN based solutions partition the key space in multiple dimensional large zones, leading to flooding at the edge once reaching the zone. The dimension conversion is very restrictive, since once set, it is hard to encode anything else apart from zones, and therefore even for exact match query a range query is actually processed at the edge [4].

For this a Hilbert-based transformation function used in this paper has the advantage of offering both the possibility to convert any multi-dimensional attribute space to a one-dimensional hash value. It also allows an exact match query as well. Knoll et al. [6] have demonstrated the clustering property of the Hilbert Curve when applied to modelling geographic coding. Moon et al. [8] have made a pioneering study on the clustering property of the Hilbert-curve. And in this paper, the same space filling curve is used as a main cornerstone to design location-based distributed overlay systems.

3 Architecture

The proposed architecture could be used depending on the location awareness capabilities at the end-device. The terminal should be able to detect its capabilities and location, and to query the P2P architecture.

The main advantage of using DHTs in P2P applications is to provide a scalable system capable of storing large numbers of items while minimizing the routing cost and offering a load-balanced indexing of entries among peers. Chord, for instance, requires each peer node to store routing information for $O(logN)$ other peers, where N is the total number of participating peers. Similarly, it takes a maximum $O(logN)$ routing messages to find any content in a Chord ring.

A peer is a computational entity (which could be a virtual one) capable of mediating information about resources (in this case wireless cells and their status) stored or managed by the underlying server infrastructure (as shown in Figure 2). These peers are named access peers and their role is organizing the underlying infrastructure and clustering the data items managed by that infrastructure.

The peers which are accessible to the mobile users are called search peers. They can play the role of a proxy to the mobile user. They initiate the range query required for supporting the context-aware vertical handover. They also manage mobility in the sense that they redirect results of the query back to the user's current address. This could be similar to the I3 based host mobility solution [16], where the user is identified through a constant overlay ID and a variable value. This tuple is called a trigger and is stored transparently in the overlay network. Messages destined to the mobile node need just to reach the overlay node that stores the trigger. The latter overlay node is

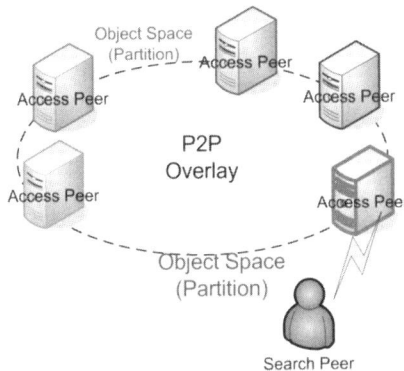

Fig. 2. P2P solution for organizing topology servers

informed by the mobile node each time the IP address or location ID is changed. This indirection process, is transparent to the correspondent nodes, differently to Mobile IP, and is highly distributed (no central home agents anymore).

Chord manages XML description files, retrievable by search peers and stored by access peers. The content or object space is no longer randomized through the use of a hash function such as *SHA-1*, distributing the keys among the peers in load-balanced way. In contrast, we rely for our system of Geocoding [11] to identify objects *(key,value)* through geographic *key* pairs among the peers. This could be compared to an online map service that can generate a zip code from entered GPS coordinates. The problem with Chord is that its query language requires a precise known object ID that can generate a precise (key, value) pair, making range queries difficult. For that reason, the ID is divided into blocks where the higher significant blocks represent unique larger areas of the earth surface, while the smaller significant blocks identify partitions of the large areas.

4 Analytic Study to Query Overhead

In order to achieve the sought distribution of data items among the peers, the following functions in a Chord like DHT need to be modified:

- Inserting an object in the overlay requires a coding function from the two-dimensional geographic coordinates to a single coordinate which preserves neighbourhood
- Query definition depends on a similar transformation function transforming a geographic range into the set of keys
- A set of neighbouring keys in the Chord space are easily converted back into a geographic range

For the above functionalities to be fulfilled, a space filling curve could be used.

4.1 Space Filling Curves

Peano curves are space filling curves used in different domains from data compression to databases. Three Peano curve examples are illustrated in Figure 3. In the figure a 2-*dimensional* finite space has been divided in a grid of *4 × 4* square cells, and then filled by three different Peano curves. Each cell could be discretely addressed by x_n- and y_n-coordinates (e.g. the top left cells, covered by the query box, have the addresses *(0,1), (0,2), (0,3),(1,1),(1,2),(1,3)*). The Peano curve connects the grid cells with a directed line, which indicates the order of transition between the cells. Assuming that each cell is assigned an integer identifier, incrementally increasing following the shape and direction of the curve, the resulting new addresses are one-dimensional. Each 2-dimensional discrete coordinate pair (x_n, y_n) corresponds to a single integer ID. In general, a space filling curve is said to map a *d*-dimensional coordinate system (in the range $[0, 2^{\gamma} - 1]^d$) to a one-dimensional coordinate system ranging $[0, 2^{\gamma \cdot d} - 1]$, where γ is also known as the degree of the space filling curve. The degree of a curve relates to the size of the underlying *d*-dimensional grid. In the 2-dimensional case, the filled space is a square surface divided in $2^{\gamma} \times 2^{\gamma}$ grid cells.

All Peano curves, including the Hilbert curve, have a fractal nature in the way they could be extended to describe the same space in a higher granularity. This depends on the degree of the space filling curve. In Figure 4, four examples are illustrated for

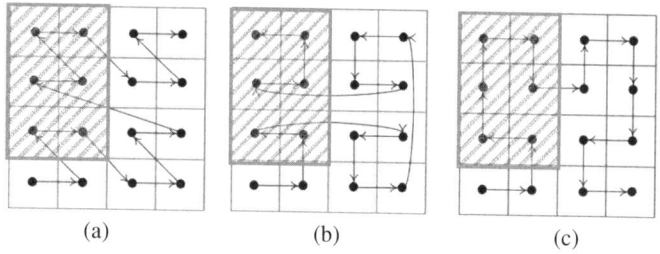

| (a) | (b) | (c) |

Fig. 3. 2-dimensional Peano space-filling curves: (a) z-curve, (b) Gray-coded curve, (c) Hilbert curve

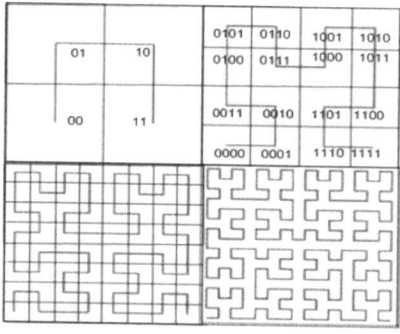

Fig. 4. $\gamma = 1, 2, 3, 4$ illustrations ($H_1^2, H_2^2, H_3^2, H_4^2$)

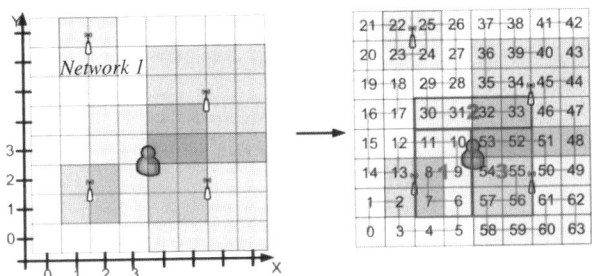

Fig. 5. Hilbert curve-based transformation from $(x,y)^{64}$ coordinates to 64 one-dimensional 6-Bit IDs ($d=2$ *and* $\gamma = 3$)

degrees γ ranging from 1 to 4. The degree of the space filling curve affects the number of integers assigned to each cell. When represented in binary, this ID has to be $2 \cdot \gamma$-bits long.

Knoll et al. [6] have shown that out of the Peano space filling curves, the Hilbert curve maps geographic neighborhood best. In other words, given two neighboring binary addresses obtained following a Hilbert curve, these two points are geographically closest compared to other space filling curves. This property is also called clustering. "A Cluster is a group of grid points inside the query that are consecutively connected by a mapping (or a curve)" [8]. In Figure 3, the Hilbert curve already shows that there is a single cluster covered by the query box (the 6 cells in the top left corner of each grid). On the other hand, the Gray-coded curve and Z-curve require two clusters for the same range box. It is shown in [8] that, in a 2-dimensional space, a range query is best described as a rectangular area. This self-organizing aspect allows any geographically close objects to be clustered together fulfilling our requirement for preserving the semantic neighborhood relation of the stored data items among our peers.

The example shown in Figure 5 illustrates the transformation of a 64-cell grid space addressed with (x_n, y_n) coordinates, where $(x_n, y_n) \in N^2 \mid 0 \leq x_n \leq 7$, *and* $0 \leq y_n \leq 7$, As a result is $2^3 \times 2^3$ one-dimensional ID space requiring $3+3 = 6$ bits. In the shown example, there are *4* access cells requiring each a different number of keys to describe their geographic coverage. As an example *"Network 1"* (in Figure 5) requires four keys *22, 24, 25, 26*. The clustering effect is demonstrated by the given query box in the middle. To start the query, the start of the cluster and its length need to be identified. Here, three clusters exist (*[6,11], [30, 33]*, and *[52,57]*), requiring each a separate get(key) call. Identifying the peers responsible of each cluster is bounded by the overlay routing cost *log(Np)* (where *Np* is the number of peers).

4.2 Effect of Transformation Function on Query Complexity

In this subsection the complexity of a geographic query is estimated analytically. The result should indicate the implementation parameters that need to be optimized in order to model the earth surface as a whole then, more specifically, an urban area, and finally some heterogeneous wireless cells. Recall that a range query is carried out as follows:

- First the clusters within the given query box need to be identified
- A *get(key)* message is sent for to reach the peer responsible of the first key within that cluster iteratively
- Hopping between peers covering the cluster follows recursively

Overhead estimation of the search process
To illustrate that, Figure 6 shows how the search for a single cluster is done (e.g. cluster 1: [6,11]). In total, *O(log(Np)+M)* messages are required.

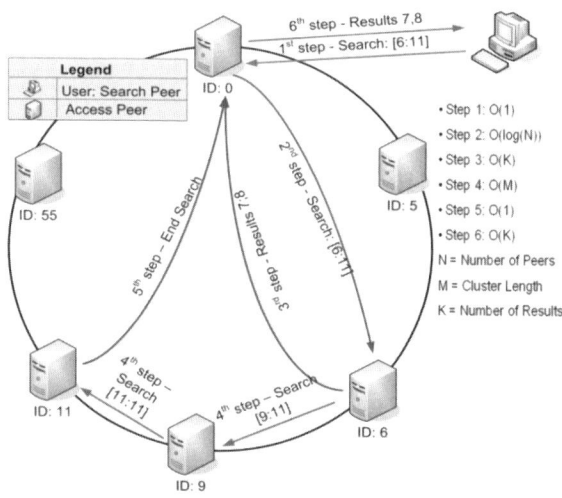

Fig. 6. Search for cluster 1 (from ID 6 to ID 11) shown in Figure 5

Table 1. List of variables used in the asymptotic study

Variables	Definition
$\gamma = k+n$	Binary index of a side distance of a 2-dimensional grid space (2^{k+n} unit for the maximum x-distance and 2^{k+n} unit for the maximum y-distance)
k	Binary index of a square search window, i.e. an area of $2^k \times 2^k$
H^2_{k+n}	A 2-dimensional Hilbert Curve filling a Geographic grid space with an area $2^{k+n} \times 2^{k+n}$
$N_2(k,k+n)$	The average number of clusters within a query box of size $2^k \times 2^k$ in a $2^{k+n} \times 2^{k+n}$ region filled by a H^2_{k+n} Hilbert Curve
M	The number of peers covering a single cluster (worse case)
M'	The average number of peers storing keys for a single cluster (simulated partition of the key space per peer)
Np	Total number of peers taking part in the Chord network
Np'	Number of peers storing data items for a portion of the grid space, e.g. a city

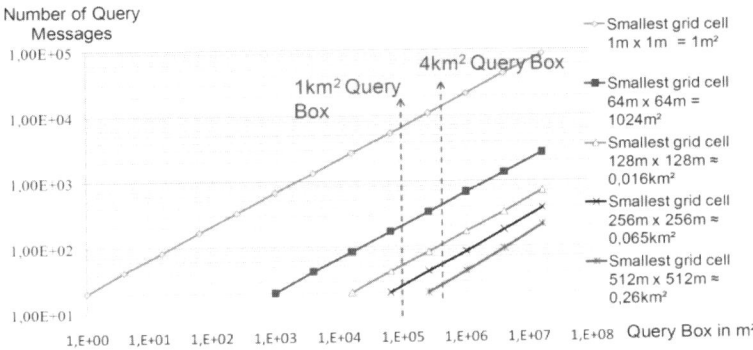

Fig. 7. Asymptotic varying query box for different Hilbert granularities (log scale)

The average number of clusters within a query box of size $2^k \times 2^k$ in a $2^{k+n} \times 2^{k+n}$ region, covered by a H_{k+n}^2 Hilbert curve is given in [8] as $N_2(k, k+n)$:

$$N_2(k, k+n) = \frac{(2^n-1)^2 2^{3k} + (2^n-1)2^{2k} + 2^n}{(2^{k+n} - 2^k + 1)^2} \quad (1)$$

The value of M indicates the fragmentation of a cluster among neighboring peers (two peers in Figure 6). For worst case estimation, we assume that each key is located at a different peer. The value of M can then be estimated as the length of each cluster.

$$M(k, k+n) = \frac{number\ of\ keys\ in\ a\ query\ box}{avg\ number\ of\ clusters\ within\ that\ box} - first\ hop$$

Or:

$$M(k, k+n) = \frac{2^k \times 2^k}{N_2(k, k+n)} - 1$$

The resulting complexity of the query box of size $2^k \times 2^k$ in space by a 2-dimentional Hilbert curve of degree ($\gamma = k+n$) is given in Equation (2).

$$O(N_2(k, k+n) \cdot \log_2(N_p) + 2^{2k} - N_2(k, k+n)) \quad (2)$$

Analytic optimization of query overhead

Figure 7 shows the effect of choosing the right Hilbert curve degree on the overhead of a range query. The earth surface could be modeled with different cell-sizes translating in different Hilbert curve degrees $(k+n)$. The cases shown in Figure 7 are those of $(k+n) = 19, 18, 17, 16$, resulting in a Chord key space *of 38, 36, 34, and 32* bits respectively.

The cell size of each grid cell is given in the figure. The query box (see Table 2) is varied from $k = 0$ to $k = 12$, resulting in a different geographic range depending on the grid size used. The more granular the Hilbert curve is, that the larger the overhead

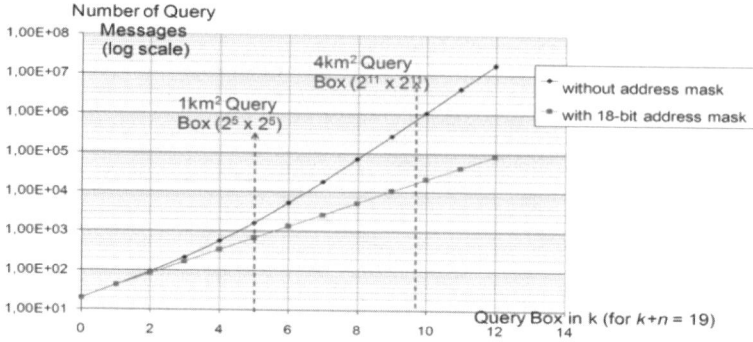

Fig. 8. Search complexity worst case scenario ($k+n = 19$, $m=20$), 18-bit peer ID masking

Table 2. Analytic evaluation of complexity for different Hilbert curve degrees

Parameter	Geographic Interpretation	Chord Interpretation
Earth Surface	300 000 km²	Total modeled surface
$2^{k+n} \times 2^{k+n}$	Number of cells in a grid covering the earth surface the size of each cell is $cell\_size = \dfrac{3 \times 10^{11}}{2^{2(k+n)}} m^2$	Hilbert Curve H_{k+n}^2 the Chord key length is $(2 \cdot (k+n))$-bit long
$Np = 2^d$	$2^{2(k+n)}$ grid cells through 2^d peers	Each peer manages $2^{2(k+n)-m}$ possible keys, where $2 \cdot (k+n) > m$
$m = 20$	The surface covered by each peer is $\dfrac{3 \times 10^{11}}{2^{20}} \approx 3 \times 10^9\, m^2$	At least a H_{10}^2 or each peer has a 20-bit long Chord ID, the remaining $(2 \cdot (k+n) - 20)$ bits could be masked
Query Box $2^k \times 2^k$	Objects: Depending the modeled earth grid, each query box covers $2^{2k} \times cell\_size$	Each query box requires $N_2(k, k+n)$ searches
Query Box $2^k \times 2^k$	Peers: number of involved access peers in the search is $\left\lceil \dfrac{2^{2k}}{2^{2(k+n)-m}} \right\rceil$	Each cluster requires M' recursive hops

becomes. This is partly due to the fact that each object (in our case a wireless cell) of let say *500m × 500m* size requires *2500* Chord keys when using $\gamma = k+n = 19$, and only *one* Chord key when using $\gamma = k+n=15$. The problem with $\gamma =15$, though, is that a smaller cell of let say *50m × 50m* will be encoded as well in a *500m × 500m* cell adding a significant loss of information about the stored data in the Chord ring.

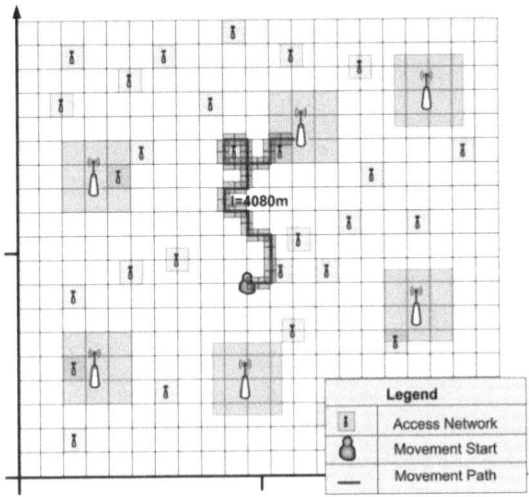

Fig. 9. A portion of the simulated environment with example query boxes in red

The masking of the peer's ID reflects the fractal nature of the Hilbert curve. In the modelled space, the earth surface is covered by a $2^{19} \times 2^{19}$ grid, which covers the earth surface with a grid cells of a little more than $2^0 \times 2^0$ $1m^2$ (taking the earth surface to be *300 000 km²*).

The effect of using a different grid cell size is shown in the Figure 7. This also corresponds mathematically to a different degree Hilbert curve and therefore to a different number of bits for the ID space (see Table 2).

Masking used in Figure 8 allows us to address peers with 20-bits instead of the full 38-bit full length IDs. Therefore the partition of the modeled clusters could be optimized. The actual length of a cluster becomes *M'* (see Table 2) where:

$$M' = \begin{cases} 1 & , in\ case\ \dfrac{2^{2k}}{N_2(k,k+n)} - 1 < 2^m \\[3mm] \dfrac{2^{2k} - N_2(k,k+n)}{N_2(k,k+n)} + 1 & ,\ otherwise \end{cases}$$

Where $2^{2(k+n)-m}$ is the minimal distance between each two Chord nodes. In our example we took *m=20*.

5 Simulation Scenarios

Java-based discrete event simulation is used to validate the analytic model. Only the overlay behavior is targeted. The user mobility and query needs differ according to the assumed scenarios. Lookup strategies are chosen according to the modeled user behavior and the required queries are generated to mainly evaluate the system.

5.1 Network Model

The simulation model assumptions are summarized in Table 3:

Table 3. Network Simulation Model

Elements	Values
Network model	Chord nodes are simulated at the overlay level, i.e., fully reliable nodes and no packetization or delays
Overlay nodes	Both access and search peers are simulated. Chord is implemented. Node IDs are obtained using the Hilbert curve partition of the key space
Simulated object space	Using a fixed grid cell size of $16 \times 16\ m^2$, i.e., 2^{38-8} possible keys or 30-bit key space ($k+n=15$)
Urban area	268 km^2 large, where access networks and users are located, e.g. Passau in Germany is about 69km^2 large
Peer distribution	2^{10+p} of which, 2^p responsible of the 268 km^2 city and the remaining 2^{10} for the rest of the earth
Chord IDs per peer	Each peer within the urban area manages $2^{2(k+n)-m}$ Chord keys and covers a geographic zone of ($16^2\ m^2 * 2^{30-p}$)
p	10, 9, 8, 6, and 4
Peer ID mask	Masking 20-bit peer IDs[2]

5.2 Object and Query Model

Given the geographic area of a medium-sized city, heterogeneous wireless cells are modeled as randomly distributed cells. A wireless cell of radius r is fitted inside a square grid cell of size $2r$ as already explained in Section 4. Similar to [15], 50 wireless cells are distributed randomly for each $1km^2$. That makes about 13400 access networks for the whole $268km^2$ modeled city (See Table 3).

Since wireless cells are heterogeneous in nature, we consider different cell sizes. Macro cells whose radius is up to 128m result in grid area of $256m \times 256m \equiv 16m \cdot 2^s \times 16m \cdot 2^s$, where $2^{2 \cdot s}$ represents the number of keys required to encode this cell in the Chord space, where "s" is varied uniformly between 1 and 4.

A graphical representation of part of the urban city is shown in Figure 9. Further to the object model, the query is also modeled according to some movement pattern assumptions.

- We assume that the user follows a random walk through a Manhattan type of street model.
- The Manhattan grid street system assumes a junction each $256m$ a randomly selected direction is taken at each junction with probabilities: 0.5 for continuing in the same direction and probability 0.25 for turning either right or left.

---

[2] Since Chord uses a fixed ID space, each peer gets a 20-bit long ID with the remaining 10 least significant bits set to zero.

- The starting point of the movement is fixed as well as the travelled distance to *4080m*.
- The velocity of the movement restricts the scope and maximum size of the query box, since the movement pattern can change by each street junction, i.e. every $(\tau = \dfrac{256m}{velocity})$ seconds.

The resulting query boxes could be said to have the following properties.

- Assuming smaller query boxes as the full length of the traversed path, query boxes are emitted by the user recursively as he/she progresses along the movement path.
- A query box cannot be generated as long as the user has not moved out the area where his previous query has been generated, this excludes for slow users to generate large queries to frequently.
- A user moving with a given linear velocity crosses each single Chord key every *16m*.
- Each side of a query box covers a distance of $2^k \cdot 16m$. Given the user requires to generate queries periodically with a period $\Delta \tau$ then the query box could be defined as follows:
- $k = \log_2 \dfrac{velocity \times \Delta \tau}{16m}$, where k is a positive integer.

Assuming three different speeds (*2m/sec, 8m/sec,* and *16m/sec*) corresponding to (*7.2km/h, 28.8km/h,* and *57.6km/h*) respectively, and different periods for generating queries $\Delta \tau = 1, 2, 4, 8, 16, 32, 64$ sec , the resulting minimum query size is given in Table 4. The resulting number of recursive query boxes for each k and each speed is listed in Table 5.

Table 4. The possible exponent k used for the query box

-	$\Delta \tau = 1$	$\Delta \tau = 2$	$\Delta \tau = 4$	$\Delta \tau = 8$	$\Delta \tau = 16$	$\Delta \tau = 32$	$\Delta \tau = 64$
Velocity = 2m/sec	-	-	-	0	1	2	3
Velocity = 8m/sec	-	0	1	2	3	4	5
Velocity = 16m/sec	0	1	2	3	4	5	6

Table 5. Number of required query boxes for the length 4080m

-	$k = 0$	$k = 1$	$k = 2$	$k = 3$	$k = 4$	$k = 5$	$k = 6$
Number of query boxes Q for *4080m*	255	128	64	32	16	8	4

5.3 Measurements

Measurements take both message counts of both query messages and those contained in the responses. The exact message size is not of interest here. It is simply assumed to be of several bytes (or *kbits*). This is an implementation issue and depends on the format of the messages.

The messages sent back as part of a response are also measured as both message units (where each unit represents an additional Chord key sent back).

It is worth noting that a peer does not send a message unit for each key found, but rather constructs an XML description file, for instance, with a list of found keys. The size of the XML file however varies for each scenario.

5.4 Simulation Results

Search Overhead vs. Query Results

For this purpose, the network topology and object models are those found in Table 3, Table 4, and in Table 5. The number of peers covering the urban zone is initially set to 2^{10}, i.e. $m=10$ in Table 3. For the same distance of *4080m,* the number of query boxes needed are given in Table 4, whereas the frequency of the queries is $\Delta\tau$.

The involved overlay messages including all the *get(key)* calls needed to cover each query box is measured. According to the Hilbert transformation of both the user's position, and the clusters included in each box, the final number of messages are summed up together in Figure 10 (a). For $k = 0$, the shown results in Figure 10 (a) are those of an exact match query occurring along the movement path and therefore repeating *255* times (see Table 5). The first range query occurs by $k = 1$ resulting in a large amount of messages.

It could be concluded that the larger the query box is that the fewer the number of messages needed to cover the same range (Figure 10-a). For each scenario a different random path has been selected indicating the slight difference that exists for each speed. Recall dependent on the size of the query box a Chord search is started for each cluster inside that box. Due to the random direction of our random path, there might be some slight differences in the number of clusters that result for each path.

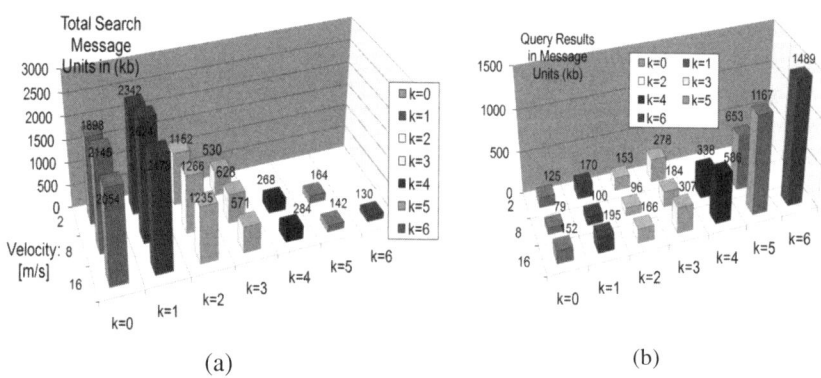

(a)	(b)

Fig. 10. Simulation results, for 4080m random path for test pair (*k, velocity*): (a) search messages overhead; (b) response messages overhead

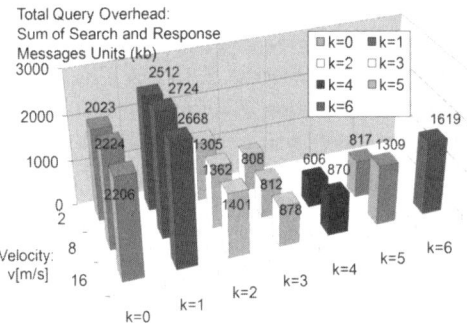

Fig. 11. Sum of search and response message overhead for each test pair (k,velocity)

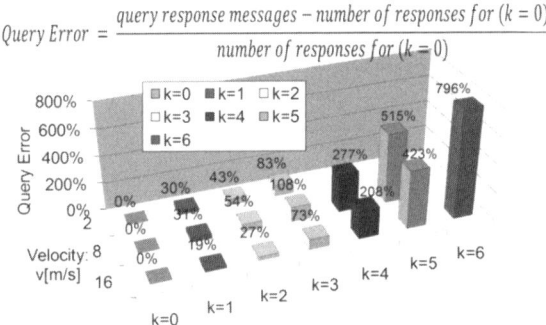

Fig. 12. Query error (%)

The amount of recorded key responses (message units), shown in Figure 10-b, increases with increasing k. The case for $k=0$ shows that the number of found keys vary according to the path taken as well as according to the object distribution. Due to the randomness of the two parameters, some difference could be noticed for each speed.

The summation of both query messages and response message units - (see Figure 11) - shows the effect the selected query box for the selected context. The ideal query size would be in that case $k = 3$ or $k = 4$. But since $k = 4$ does not suit a slow moving user with a random walk as defined in Table 4, it could be concluded that $k=3$ is the more suitable query box size extracted from all three movement velocities.

The velocity of the movement shows little effect on the message count itself, but it could be said that the overall throughput cost of the range query could be calculated as follows:

$$Query\ Throughput\ =\ \frac{query\ messages\ count \times velocity}{distance}$$

This is an indication of the scalability of the whole approach. Other mechanisms such as Caching (at the search peers for instance) could be used to re-use similar query results generated from different users. Assuming that queries are initiated simultaneously, originating from different peers, the asynchronous nature of the query definition plays a major role in limiting the traffic. A slow user generates queries at a totally different frequency to a fast user. The fast user could generate larger queries to limit the frequency of the messages sent but then would have to tolerate a given error in the response messages.

A Query Error Measure

The error of a query is given as a function of the exact match query results (i.e., with k = 0) as follows:

$$Query \ Error_k = \frac{query \ response \ count \ (k) - query \ response \ count \ (for \ k = 0)}{query \ response \ count \ (for \ k = 0)}$$

The query error mostly has a computational cost, since the user has to filter out the messages that are sent back. For the case of $k=3$ which in our scenario has produced the ideal amount of traffic messages, between 73% and 108% is noted (depending on the path selected). The trend of the query error follows that of the number of obtained results as shown in Figure 10-b. The query error is plotted for each $(k, velocity)$ pair in Figure 12.

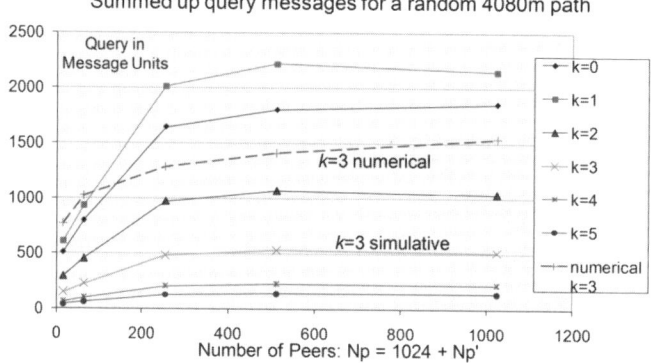

Fig. 13. The effect of number of urban peers (Np')

5.5 Comparing the Simulation Results with the Numerical Ones

In Figure 13, the number of peers which are managing the same urban zone is varied. The behaviour of the query message cost is compared. The query responses depend on the objects stored in the system and in this case this has been kept the same i.e. 13400 wireless cells managed by our urban peers.

The smallest number of urban peers $(Np'= 16)$, is the most centralized scenario. The most distributed scenario simulates $Np' = 1024$ as given in Table 3. This indicated by parameter "$p = 4, 6, 8, 9, 10$" corresponding to $Np' = 16, 64, 256, 512, 1024$).

For a smaller query boxes ($0 \leq k \leq 2$), the query overhead is mostly affected by the number of peers. For larger query boxes ($3 \leq k \leq 4$) the effect of the number of peers is minimal. Now recall our complexity equation (Equation (2) in Subsection 4.2) $O(N_2(k, k+n) \cdot \log_2(N_p) + 2^{2k} - N_2(k, k+n))$. The overhead cost is repeated for each single query box that covers the movement path. In our simulated scenario, this represents a multiplication factor Q given in Table 5. The cost complexity is then transformed to:

$$O(Q \cdot [N_2(k, k+n) \cdot \log_2(N_p) + 2^{2k} - N_2(k, k+n)])$$

Where:

$$Q = \frac{full\ length\ movement\ path}{grid\ cell\ size \times 2^k} = \frac{4080m}{16m \times 2^k}$$

Furthermore, the number of clusters $N_2(k, k+n)$ (see Equation (1), Subsection 4.1) decreases with increasing size of the query window k. This also explains the influence of the logarithmic factor ($log(Np)$) on the overhead cost. For larger k the number of clusters decreases, making the influence of $N_2(k, k+n) \times \log(Np)$ part of the equation smaller. For smaller k this part of the equation leads to a significant increase.

Another result – explained but not plotted here due to lack of space – is the number of overlay hops required to reach each cluster (i.e. log (N) in Chord). This has been measured to be around $0.5 \cdot \log(Np')^3$ instead of $\log(Np)$, where $Np= 1024+Np'$ (see Figure 13). The use of Np' could be explained by the fact that the search peer is close (on the overlay level) to the queried range. This differs to a scenario where a location-based query is started from Europe, for instance, and destined at the US (e.g. "from Passau – Europe – find all green spaces in Chicago – US"). The average routing cost is then given in Equation (3).

$$Average\ Cost = Q \cdot [N_2(k, k+n) \cdot 1/2 \cdot \log_2(N_p') + 2^{2k} - N_2(k, k+n)]) \qquad (3)$$

Based on Equation (3), the average numerical number of query messages (shown by the dashed line for $k=3$) is plotted in Figure 13. Equation (2) assumes that each key is placed at a distinct peer. Once the query reaches the cluster, M hops are needed to cover the cluster length (described by variable M in Table 1). In the simulation M' turns out to be at most one additional hop. This tendency is shown by the simulative results for ($k=3$) when compared with the numerical result (Equation 3). Both results follow the same growth tendency. The logarithmic growth (i.e., $log_2(2 \cdot Np')$ $=log_2(Np')+1$) explains the flattening of the cost after $Np'= 512$ in all cases. It is hard to predict, numerically, the splitting of a cluster between peers. In the case of $k=1$, the clusters' partition among peers has lead to a slight decrease in the search cost (from $Np' = 512$ to $Np'= 1024$).

---

[3] In Chord the average number of hops is given as $0.5*log(N)$, where N is the number of peers [14].

6 Conclusions

The proposed context-aware handover relies on application layer overlay to access information stored by heterogeneous data bases. Based on the clustering properties of geographic information, the Hilbert curve is used to obtain a transformation function between the object storage attribute (geographic zones) and their allocated ID in the overlay system. The fractal nature of the Hilbert curve is exploited to address overlay nodes while preserving the clustered structure of the managed data. Numerical analysis of the query problem is used to further analyse the complexity of the sought system. A simulation study, then, investigates the performance of the range queries and compares the results to the expected numerical cost estimation.

The proposed solution is shown to be affected by very local decisions. For instance the size of the query could be optimized locally, and the effect of the Chord complexity only takes the local urban context into account.

As a future work, further scalability studies of the system will be investigated as well as applicability of the design path to other distributed management systems.

Acknowledgements

This work has been supported by the EURO-FGI -Network of Excellence, European Commission grant IST 028022 and specifically the subprojects VNets and MMSOS. The authors would like to thank Andreas Berl for providing insightful comments to the ideas presented in this paper, and Ivan Dedinski who provided valuable support for the simulation work.

References

1. Akyildiz, I.F., Wang, X., Wang, W.: Wireless Mesh Networks: A Survey. Computer Networks Journal 47(4), 445–487 (2005)
2. Akyldiz, I.F., Xie, J., Mohanty, S.: A Survey of Mobility Management in Next Generation all-IP based Wireless Systems. IEEE Wireless Communications 11(4), 16–28 (2004)
3. Balasubramaniam, S., Indulska, J.: Vertical Handover Supporting Pervasive Computing in Future Wireless Networks. Computer Communications 27(8), 708–719 (2004)
4. Datta, A., Hauswirth, M., John, R., Schmidt, R., Aberer, K.: Range Queries in Trie-Structured Overlays. In: Proc. of Fifth Peer-to-Peer Computing Conference, August 31-September 2, 2005, pp. 57–66. IEEE Press, Konstanz, Germany (2005)
5. Houyou, A.M., De Meer, H., Esterhazy, M.: P2P-based Mobility Management for Heterogeneous Wireless Networks and Mesh Networks. In: Cesana, M., Fratta, L. (eds.) Euro-NGI 2005. LNCS, vol. 3883, pp. 226–241. Springer, Heidelberg (2006)
6. Knoll, M., Weis, T.: Optimizing Locality for Self-organizing Context-Based Systems. In: De Meer, H., Sterbenz, J.P.G. (eds.) IWSOS 2006. LNCS, vol. 4124, pp. 62–73. Springer, Heidelberg (2006)
7. Maymounkov, P., Mazieres, D.: Kademlia: A Peer-to-Peer Information System Based on the XOR Metric. In: Druschel, P., Kaashoek, M.F., Rowstron, A.I.T. (eds.) IPTPS 2002. LNCS, vol. 2429, pp. 53–65. Springer, Heidelberg (2002)

8. Moon, B., Jagadish, H.V., Faloutsos, C., Saltz, J.H.: Analysis of the Clustering Properties of the Hilbert Space-Filling Curve. In: IEEE Transactions on Knowledge and Data Engineering, vol. 13(1), pp. 124–141. IEEE Press, Los Alamitos (2001)

9. Ratnasamy, S., Francis, P., Handley, M., Karp, R., Schenker, S.: A scalable content-addressable network. In: SIGCOMM 2001: Proceedings of the 2001 conference on Applications, Technologies, Architectures, and Protocols for Computer Communications, pp. 161–172. ACM Press, New York (2001)

10. Sahin, O.D., Gupta, A., Agrawal, D., El Abbadi, A.: Query Processing Over Peer-to-Peer Data Sharing Systems. Technical Report UCSB/CSD-2002-28, University of California at Santa Barbara (2002)

11. Schiller, J., Voisard, A.: Location-Based Services. Morgen Kaufmann/Elsevier, San Fransisco (2004)

12. Sethom, K., Afifi, H., Pujolle, G.: Palma: A P2P based Architecture for Location Management. In: 7th IFIP International Conference on Mobile and Wireless Communications Networks (MWCN 2005), Marrakech, Morocco, September 19-21 (2005)

13. Stemm, M., Katz, R.H.: Vertical Handoffs in Wireless Overlay Networks. Mobile Networks and Applications 3(4), 335–350 (1998)

14. Stoica, I., Morris, R., Liben-Nowell, D., Karger, D.R., Frans Kaashoek, M., Dabek, F., Balakrishnan, H.: Chord: a Scalable Peer-to-Peer Lookup Protocol for Internet Applications. IEEE/ACM Transactions on Networking 11(1), 17–32 (2003)

15. Zdarsky, F.A., Martinovic, I., Schmitt, J.B.: The Case for Virtualized Wireless Access Networks. In: De Meer, H., Sterbenz, J.P.G. (eds.) IWSOS 2006. LNCS, vol. 4124, pp. 90–104. Springer, Heidelberg (2006)

16. Zhuang, S., Lai, K., Stoica, I., Katz, R., Shenker, S.: Host Mobility Using an Internet Indirection Infrastructure. Wireless Networks 11(6), 741–756 (2005)

Author Index